Fire Protection, Detection, and Suppression Systems

Fifth Edition

Cindy Brakhage
Project Manager/Senior Editor

Alex Abrams and Jeff Fortney
Senior Editors

Brad McLelland and Lindsey Dugan
Lead Instructional Developers

P9-CRU-827

IFSTA

Validated by the International Fire Service Training Association

Published by
Fire Protection Publications • Oklahoma State University

RECYCLABLE

The International Fire Service Training Association (IFSTA) was established in 1934 as a *nonprofit educational association of fire fighting personnel who are dedicated to upgrading fire fighting techniques and safety through training.* To carry out the mission of IFSTA, Fire Protection Publications was established as an entity of Oklahoma State University. Fire Protection Publications' primary function is to publish and distribute training materials as proposed, developed, and validated by IFSTA. As a secondary function, Fire Protection Publications researches, acquires, produces, and markets high-quality learning and teaching aids consistent with IFSTA's mission.

IFSTA holds two meetings each year: the Winter Meeting in January and the Annual Validation Conference in July. During these meetings, committees of technical experts review draft materials and ensure that the professional qualifications of the National Fire Protection Association® standards are met. These conferences bring together individuals from several related and allied fields, such as:

- Key fire department executives, training officers, and personnel
- Educators from colleges and universities
- Representatives from governmental agencies
- Delegates of firefighter associations and industrial organizations

Committee members are not paid nor are they reimbursed for their expenses by IFSTA or Fire Protection Publications. They participate because of a commitment to the fire service and its future through training. Being on a committee is prestigious in the fire service community, and committee members are acknowledged leaders in their fields. This unique feature provides a close relationship between IFSTA and the fire service community.

IFSTA manuals have been adopted as the official teaching texts of many states and provinces of North America as well as numerous U.S. and Canadian government agencies. Besides the NFPA® requirements, IFSTA manuals are also written to meet the Fire and Emergency Services Higher Education (FESHE) course requirements. A number of the manuals have been translated into other languages to provide training for fire and emergency service personnel in Canada, Mexico, and outside of North America.

ISBN 978-0-87939-599-5   Library of Congress Control Number: 2016934558

Fifth Edition, First Printing, April 2016   *Printed in the United States of America*

10   9   8   7   6   5   4   3   2

If you need additional information concerning the International Fire Service Training Association (IFSTA) or Fire Protection Publications, contact:

Customer Service, Fire Protection Publications, Oklahoma State University
930 North Willis, Stillwater, OK 74078-8045
800-654-4055   Fax: 405-744-8204

For assistance with training materials, to recommend material for inclusion in an IFSTA manual, or to ask questions or comment on manual content, contact:

Editorial Department, Fire Protection Publications, Oklahoma State University
930 North Willis, Stillwater, OK 74078-8045
405-744-4111   Fax: 405-744-4112   E-mail: editors@osufpp.org

Chapter Summary

Table of Contents

List of Tables

Acknowledgements

The fifth edition of the **Fire Protection, Detection, and Suppression Systems** (previously titled **Fire Detection and Suppression Systems, Fourth Edition**) is intended to be an introductory learning resource for those individuals who may be responsible for the design, installation, inspection, and maintenance of these systems. In addition, the manual is also intended to be a resource for emergency services personnel who may respond to incidents in protected premises. The manual has been revised to reflect the most current equipment, technology, and practices in the field.

Acknowledgement and special thanks are extended to the members of the IFSTA validation committee that contributed its time, wisdom, and knowledge to the development of this manual.

IFSTA Fire Protection, Detection, and Suppression Systems
Fifth Edition Validation Committee

Chair
Josh M. Stefancic
Division Chief
Largo Fire Rescue
Largo, FL

Secretary
Dan Ripley
Fire Captain
Lincoln Fire and Rescue
Lincoln, NE

Committee Members

Shane Alexander
Battalion Chief
Ocala Fire Rescue
Ocala, FL

Brad Austin, PE, CSP
Fire Protection Engineer/Firefighter-EMT
Poole Fire Protection, Inc./Derby (KS) Fire and
 Rescue Department
Olathe, KS

Russell B. Bainbridge III, PE, CFPS
Engineering Specialist
CNS Pantex Plant
Amarillo, TX

Andrew Barr, CFPS
Deputy Fire Marshal
McKinney Fire Department/Fire Marshal's
 Office
McKinney, TX

Pat Brock
Professor Emeritus
Fire Protection and Safety Technology
Oklahoma State University
Stillwater, OK

Christopher L. Conroy, CET
Senior Fire Protection Specialist
GHD
Chantilly, VA

John S. Cunningham
Executive Director
Nova Scotia Firefighters School
Waverley, NS

Tonya L. Hoover
State Fire Marshal
CAL FIRE
Sacramento, CA

Bill Longworth, EMT-P, CET, CFPS
Fire Suppression Systems and Outage
 Coordinator
Battalion Chief, Oak Ridge National Laboratory
 Fire Department
Oak Ridge, TN

Jonathan Lund
Fire Marshal
Des Moines Fire Department
Des Moines, IA

IFSTA Fire Protection, Detection, and Suppression Systems
Fifth Edition Validation Committee
Committee Members (cont.)

Corey Matthews
Captain II
Fairfax County Fire and Rescue
Fairfax, VA

J. Scott Mitchell, PE
Fire Protection Engineer

Tom Mooney, IAAI-FIT
Deputy Fire Marshal
Tualatin Valley Fire and Rescue
Tigard, OR

Jimbo Schifiliti
COB
Fire Safety Consultants, Inc.
Elgin, IL

R. Paul Valentine
Fire Protection Division Manager
Jensen-Hughes
Oakbrook Terrace, IL

Much appreciation is given to the following individuals and organizations for contributing information, photographs, and technical assistance instrumental in the development of this manual:

Asco Aerospace USA
Stillwater, OK

Bert Cooper Engineering Lab
Oklahoma State University
Stillwater, OK

Chris Mickal
New Orleans, (LA) Fire Department

Consolidated Nuclear Services
Amarillo, TX

Floyd Luinstra
School of Fire Protection and Safety Technology
Oklahoma State University
Stillwater, OK

McKinney (TX) Fire Department

Ron Jeffers

Rich Mahaney

Ron Moore
McKinney (TX) Fire Department

Scott Stookey
Austin (TX) Fire Department

Sprint Center
Kansas City, MO

Oak Ridge National Laboratory, U.S. Department
of Energy
Oak Ridge, TN

Sand Springs (OK) Fire Department

Tarpon Springs (FL) Fire Rescue

Texas Fire Prevention Specialists

Tim Frankenberg, CT, CFPS
Fire Product Manager
Potter Electric Signal Company
St. Louis, MO

Tom Hughes

Underwriters Laboratories, Inc.

U.S. Navy

Last, but certainly not least, gratitude is extended to the following members of the Fire Protection Publications staff whose contributions made the final publication of this manual possible.

Fire Protection, Detection, and Suppression Systems Fifth Edition, Project Team

Project Manager
Cindy Brakhage, Senior Editor

Director of Fire Protection Publications
Craig Hannan

Curriculum Manager
Leslie Miller

Editorial Manager
Clint Clausing

Production Coordinator
Ann Moffat

Editor(s)
Alex Abrams, Senior Editor
Jeff Fortney, Senior Editor
Tony Peters, Senior Editor

Illustrators and Layout Designer(s)
Missy Hannan, Senior Graphic Designer

Curriculum Development
Angel Muzik, Instructional Developer
Brad McLelland, Instructional Developer
Crystal Griggs, Instructional Developer
Frank Carter, Instructional Developer
Jayne Ann Williamson, Instructional Developer
Lindsey Dugan, Instructional Developer
Tara Moore, Instructional Developer

Photographer(s)
Jeff Fortney, Senior Editor
Veronica Smith, Senior Editor

Technical Reviewer
Marty King
Assistant Chief
Fire Prevention and Urban Affairs
West Allis, WI

Technical Editor
Brent Meisenheimer
Lieutenant
Austin (TX) Fire Department

Editorial Staff
Tara Gladden, Editorial Assistant

Indexer
Nancy Kopper

The IFSTA Executive Board at the time of validation of the **Fire Protection, Detection, and Suppression Systems** was as follows:

Dedication

This manual is dedicated to the men and women who hold devotion to duty above personal risk, who count on sincerity of service above personal comfort and convenience, who strive unceasingly to find better and safer ways of protecting lives, homes, and property of their fellow citizens from the ravages of fire, medical emergencies, and other disasters

...The Firefighters of All Nations.

Introduction

Introduction Contents

Introduction

Courtesy of Chris Mickal/District Chief, New Orleans (LA) FD.

Fire protection, detection, and suppression systems are installed and used in a multitude of different occupancies. These systems are designed to detect the presence of fire or products of combustion, alert occupants and fire department personnel of the condition, and suppress or extinguish the fire. Components of these systems are constantly changing and improving due to advances in technology.

The fifth edition of IFSTA's **Fire Protection, Detection, and Suppression Systems** manual (previously titled **Fire Detection and Suppression Systems, Fourth Edition**) is intended to be an introductory learning resource for those individuals who may be responsible for the design, installation, inspection, and maintenance of these systems. In addition, the manual is also intended to be a resource for emergency services personnel who may respond to incidents in protected premises. The manual has been revised to reflect the most current equipment, technology, and practices in the field.

Purpose and Scope

The purpose of **Fire Protection, Detection, and Suppression Systems**, Fifth Edition, is to familiarize fire service members and other interested personnel with the components, design, maintenance, operation, testing, and inspection of common fire protection, detection, and suppression systems. This manual is not intended to substitute for training as a fire inspector or for personnel who design, maintain, and inspect such systems; however, it can serve as a valuable resource for personnel engaged in these activities. The audience for this manual includes:

- College students
- Fire service personnel
- Industrial fire protection personnel
- Building construction personnel
- All others seeking information on fire protection, detection, and suppression systems

The scope of this manual is to provide up-to-date information on fire protection, detection, and suppression systems. The manual is written based upon the current course objectives, outcomes, and outline for the associate's level course *Fire Protection Systems* as described in the *Fire and Emergency Services for Higher Education* (FESHE) core curriculum.

Book Organization

Each chapter begins with a list of Chapter Contents, Key Terms with page numbers, FESHE Outcomes, and Chapter Learning Objectives. A Case History, Key Terms with definitions, a Chapter Summary, and Review Questions are included in each chapter. Chapters include:

Chapter 1   Overview of Fire Protection, Detection, and Suppression Systems

Chapter 2   Fire Detection and Alarm Systems

Chapter 3   Water Supply Systems

Chapter 4   Water-Based Fire Suppression Systems

Chapter 5   Standpipe and Hose Systems

Chapter 6   Fire Pumps

Chapter 7   Non-Water-Based Fire Suppression Systems

Chapter 8   Smoke Management Systems

Chapter 9   Portable Fire Extinguishers

The final parts of this manual include an Appendix, a Glossary, and an Index.

Terminology

This manual is written with a global, international audience in mind. For this reason, it often uses general descriptive language in place of regional- or agency-specific terminology (often referred to as *jargon*). Additionally, in order to keep sentences uncluttered and easy to read, the word *state* is used to represent both state and provincial level governments (and their equivalent). This usage is applied to this manual for the purposes of brevity and is not intended to address or show preference for only one nation's method of identifying regional governments within its borders.

The glossary at the end of the manual will assist the reader in understanding words that may not have their roots in the fire and emergency services. The IFSTA **Fire Service Orientation and Terminology** manual is the source for the definitions of fire and emergency services-related terms in this manual.

Key Information

Various types of information in this manual are given in shaded boxes marked by symbols or icons. See the following definitions:

Case History

A case history analyzes an event. It can describe its development, action taken, investigation results, and lessons learned. Illustrations can be included.

Case History

Upon the fire crews' arrival, firefighters discovered a large fire rapidly progressing up the exterior of the building. Fire crews made an attack on the fire and quickly extinguished the flames. The outcome might have been vastly different without the presence of a fire alarm system within this building. The fire alarm system greatly contributed to the detection of the fire and notification of the buildings occupants.

Information Box

Information boxes give facts that are complete in themselves but belong with the text discussion. It is information that needs more emphasis or separation. (In the text, the title of information boxes will change to reflect the content.)

Dry Chemical Systems for Protection of Kitchen Equipment

Most existing dry chemical systems installed to protect kitchen equipment do not meet listing requirements (ANSI/UL 300, Fire Testing of Fire Extinguishing Systems for Protection of Commercial Cooking Equipment) for dry chemical fire suppression systems for commercial kitchen hood, duct, and cooking appliances. Failure to meet the listing requirements required by NFPA® 17 makes repair, annual certification, and testing difficult. This failure results in requiring the dry chemical systems protecting kitchen equipment to be removed from service and replaced with a compliant system.

Key Term

A **key term** is designed to emphasize key concepts, technical terms, or ideas that the reader needs to know. Key terms are listed at the beginning of each chapter, and the definition is placed in the margin for easy reference. An example of a key term:

Alarm Signal — Signal given by a fire detection and alarm system when there is a fire condition detected.

Signal Words

Three key signal words are found in the text: **WARNING, CAUTION,** and **NOTE.** Definitions and examples of each are as follows:

- **WARNING** indicates information that could result in death or serious injury to fire and emergency services personnel. See the following example:

> ### WARNING!
> Antifreeze is combustible. Use only listed solutions premixed at the factory.

- **CAUTION** indicates important information or data that fire and emergency service responders need to be aware of in order to perform their duties safely. See the following example:

CAUTION
Do not inhale the dry chemical agent that is dispersed into the atmosphere as it will irritate the respiratory system.

- **NOTE** indicates important operational information that helps explain why a particular recommendation is given or describes optional methods for certain procedures. See the following example:

  NOTE: When inspecting the sprinklers, a clearance of at least 18 inches (450 mm) should be maintained under sprinklers.

Metric Conversions

Throughout this manual, U.S. units of measure are converted to metric units for the convenience of our international readers. Be advised that we use the Canadian metric system. It is very similar to the Standard International system, but may have some variation.

We adhere to the following guidelines for metric conversions in this manual:

- Metric conversions are approximated unless the number is used in mathematical equations.
- Centimeters are not used because they are not part of the Canadian metric standard.
- Exact conversions are used when an exact number is necessary such as in construction measurements or hydraulic calculations.
- Set values such as hose diameter, ladder length, and nozzle size use their Canadian counterpart naming conventions and are not mathematically calculated. For example, 1½ inch hose is referred to as 38 mm hose.

The following two tables provide detailed information on IFSTA's conversion conventions. The first table includes examples of our conversion factors for a number of measurements used in the fire service. The second shows examples of exact conversions beside the approximated measurements you will see in this manual.

U.S. to Canadian Measurement Conversion

Measurements	Customary (U.S.)	Metric (Canada)	Conversion Factor
Length/Distance	Inch (in) Foot (ft) [3 or less feet] Foot (ft) [3 or more feet] Mile (mi)	Millimeter (mm) Millimeter (mm) Meter (m) Kilometer (km)	1 in = 25 mm 1 ft = 300 mm 1 ft = 0.3 m 1 mi = 1.6 km
Area	Square Foot (ft^2) Square Mile (mi^2)	Square Meter (m^2) Square Kilometer (km^2)	1 ft^2 = 0.09 m^2 1 mi^2 = 2.6 km^2
Mass/Weight	Dry Ounce (oz) Pound (lb) Ton (T)	gram Kilogram (kg) Ton (T)	1 oz = 28 g 1 lb = 0.5 kg 1 T = 0.9 T
Volume	Cubic Foot (ft^3) Fluid Ounce (fl oz) Quart (qt) Gallon (gal)	Cubic Meter (m^3) Milliliter (mL) Liter (L) Liter (L)	1 ft^3 = 0.03 m^3 1 fl oz = 30 mL 1 qt = 1 L 1 gal = 4 L
Flow	Gallons per Minute (gpm) Cubic Foot per Minute (ft^3/min)	Liters per Minute (L/min) Cubic Meter per Minute (m^3/min)	1 gpm = 4 L/min 1 ft^3/min = 0.03 m^3/min
Flow per Area	Gallons per Minute per Square Foot (gpm/ft^2)	Liters per Square Meters Minute (L/(m^2.min))	1 gpm/ft^2 = 40 L/(m^2.min)
Pressure	Pounds per Square Inch (psi) Pounds per Square Foot (psf) Inches of Mercury (in Hg)	Kilopascal (kPa) Kilopascal (kPa) Kilopascal (kPa)	1 psi = 7 kPa 1 psf = .05 kPa 1 in Hg = 3.4 kPa
Speed/Velocity	Miles per Hour (mph) Feet per Second (ft/sec)	Kilometers per Hour (km/h) Meter per Second (m/s)	1 mph = 1.6 km/h 1 ft/sec = 0.3 m/s
Heat	British Thermal Unit (Btu)	Kilojoule (kJ)	1 Btu = 1 kJ
Heat Flow	British Thermal Unit per Minute (BTU/min)	watt (W)	1 Btu/min = 18 W
Density	Pound per Cubic Foot (lb/ft^3)	Kilogram per Cubic Meter (kg/m^3)	1 lb/ft^3 = 16 kg/m^3
Force	Pound-Force (lbf)	Newton (N)	1 lbf = 0.5 N
Torque	Pound-Force Foot (lbf ft)	Newton Meter (N.m)	1 lbf ft = 1.4 N.m
Dynamic Viscosity	Pound per Foot-Second (lb/ft.s)	Pascal Second (Pa.s)	1 lb/ft.s = 1.5 Pa.s
Surface Tension	Pound per Foot (lb/ft)	Newton per Meter (N/m)	1 lb/ft = 15 N/m

Conversion and Approximation Examples

Measurement	U.S. Unit	Conversion Factor	Exact S.I. Unit	Rounded S.I. Unit
Length/Distance	10 in	1 in = 25 mm	250 mm	250 mm
	25 in	1 in = 25 mm	625 mm	625 mm
	2 ft	1 in = 25 mm	600 mm	600 mm
	17 ft	1 ft = 0.3 m	5.1 m	5 m
	3 mi	1 mi = 1.6 km	4.8 km	5 km
	10 mi	1 mi = 1.6 km	16 km	16 km
Area	36 ft²	1 ft² = 0.09 m²	3.24 m²	3 m²
	300 ft²	1 ft² = 0.09 m²	27 m²	30 m²
	5 mi²	1 mi² = 2.6 km²	13 km²	13 km²
	14 mi²	1 mi² = 2.6 km²	36.4 km²	35 km²
Mass/Weight	16 oz	1 oz = 28 g	448 g	450 g
	20 oz	1 oz = 28 g	560 g	560 g
	3.75 lb	1 lb = 0.5 kg	1.875 kg	2 kg
	2,000 lb	1 lb = 0.5 kg	1 000 kg	1 000 kg
	1 T	1 T = 0.9 T	900 kg	900 kg
	2.5 T	1 T = 0.9 T	2.25 T	2 T
Volume	55 ft³	1 ft³ = 0.03 m³	1.65 m³	1.5 m³
	2,000 ft³	1 ft³ = 0.03 m³	60 m³	60 m³
	8 fl oz	1 fl oz = 30 mL	240 mL	240 mL
	20 fl oz	1 fl oz = 30 mL	600 mL	600 mL
	10 qt	1 qt = 1 L	10 L	10 L
	22 gal	1 gal = 4 L	88 L	90 L
	500 gal	1 gal = 4 L	2 000 L	2 000 L
Flow	100 gpm	1 gpm = 4 L/min	400 L/min	400 L/min
	500 gpm	1 gpm = 4 L/min	2 000 L/min	2 000 L/min
	16 ft³/min	1 ft³/min = 0.03 m³/min	0.48 m³/min	0.5 m³/min
	200 ft³/min	1 ft³/min = 0.03 m³/min	6 m³/min	6 m³/min
Flow per Area	50 gpm/ft²	1 gpm/ft² = 40 L/(m².min)	2 000 L/(m².min)	2 000 L/(m².min)
	326 gpm/ft²	1 gpm/ft² = 40 L/(m².min)	13 040 L/(m².min)	13 000L/(m².min)
Pressure	100 psi	1 psi = 7 kPa	700 kPa	700 kPa
	175 psi	1 psi = 7 kPa	1225 kPa	1 200 kPa
	526 psf	1 psf = 0.05 kPa	26.3 kPa	25 kPa
	12,000 psf	1 psf = 0.05 kPa	600 kPa	600 kPa
	5 psi in Hg	1 psi = 3.4 kPa	17 kPa	17 kPa
	20 psi in Hg	1 psi = 3.4 kPa	68 kPa	70 kPa
Speed/Velocity	20 mph	1 mph = 1.6 km/h	32 km/h	30 km/h
	35 mph	1 mph = 1.6 km/h	56 km/h	55 km/h
	10 ft/sec	1 ft/sec = 0.3 m/s	3 m/s	3 m/s
	50 ft/sec	1 ft/sec = 0.3 m/s	15 m/s	15 m/s
Heat	1200 Btu	1 Btu = 1 kJ	1 200 kJ	1 200 kJ
Heat Flow	5 BTU/min	1 Btu/min = 18 W	90 W	90 W
	400 BTU/min	1 Btu/min = 18 W	7 200 W	7 200 W
Density	5 lb/ft³	1 lb/ft³ = 16 kg/m³	80 kg/m³	80 kg/m³
	48 lb/ft³	1 lb/ft³ = 16 kg/m³	768 kg/m³	770 kg/m³
Force	10 lbf	1 lbf = 0.5 N	5 N	5 N
	1,500 lbf	1 lbf = 0.5 N	750 N	750 N
Torque	100	1 lbf ft = 1.4 N.m	140 N.m	140 N.m
	500	1 lbf ft = 1.4 N.m	700 N.m	700 N.m
Dynamic Viscosity	20 lb/ft.s	1 lb/ft.s = 1.5 Pa.s	30 Pa.s	30 Pa.s
	35 lb/ft.s	1 lb/ft.s = 1.5 Pa.s	52.5 Pa.s	50 Pa.s
Surface Tension	6.5 lb/ft	1 lb/ft = 15 N/m	97.5 N/m	100 N/m
	10 lb/ft	1 lb/ft = 15 N/m	150 N/m	150 N/m

Overview of Fire Protection, Detection, and Suppression Systems

Chapter Contents

chapter 1

Key Terms

FESHE Outcomes

Fire and Emergency Services Higher Education (FESHE) Outcomes: Fire Protection Systems

1. Explain the benefits of fire protection systems in various types of structures.

5. Review residential and commercial sprinkler legislation.

Overview of Fire Protection, Detection, and Suppression Systems

Learning Objectives

After reading this chapter, students will be able to:

1. Describe the types of fire protection systems.
2. Summarize the historical development of fire protection, detection, and suppression systems.
3. Describe the four basic categories of fire protection, detection, and suppression systems.
4. Explain the role of standards and codes for fire detection and suppression systems.
5. Describe the benefits that fire protection, detection, and suppression systems have on life safety and loss prevention.

Chapter 1
Overview of Fire Protection, Detection, and Suppression Systems

Case History

In 2011, fire crews responded to a large warehouse fire in Texas. The warehouse, which contained high-piled storage, was originally equipped with an Early Suppression Fast Response (ESFR) sprinkler system. However, modifications were made over time. Now, extra-large orifice heads replaced ESFR heads, resulting in nearly half the flow and pressure as engineered. Suppression crews were unable to locate and contain the fire due to smoke conditions inside the structure. As the fire grew, reaching the seat of the fire became more difficult due to the waves of debris. Attempts to ventilate the structure actually contributed to the spread of the fire and possibly hindered effective sprinkler system operations instead of helping with visibility. Detection and suppression systems are only effective when proper storage height, storage arrangements, and commodity classifications are implemented in large-scale warehouses. In this case, pallets were stacked too high, preventing proper sprinkler operation. The makeup of high-rack storage material was flammable and not easily extinguished with water. When considering suppression systems, it is paramount to keep in mind the difference between control and extinguishment.

Fire prevention has long been recognized as a responsibility of fire and emergency services. In its truest form, fire prevention is the elimination of all hostile fires through education, engineering, and enforcement. It can also involve the **mitigation** of loss or limiting property loss or injury. This task can be accomplished through product regulation, design, or redesign.

Fire prevention also involves behavior modification that prevents or mitigates the threat from fires. Changing behavior can be accomplished through education and by specifically targeting high-risk audiences. Recognizing that fire prevention efforts will never be entirely effective, fire protection, detection, and suppression systems are designed to mitigate damage to both life and property when fires occur.

Fire protection, detection, and suppression systems can be part of the built environment, acting automatically upon discovery of a fire. They can also be used manually, as with fire extinguishers or manual alarm systems. This chapter serves as an introduction to fire detection and suppression systems,

> **Mitigate** — To cause to become less harsh or hostile; to make less severe, intense, or painful; to alleviate.

including the types, history, and benefits of systems. In addition, the chapter explains important codes and standards that impact the installation and use of fire protection, detection, and suppression systems.

Types of Fire Protection Systems

Passive fire protection can be a part of building construction such as a firewall, fire doors, or sprayed-on structural fire protection (**Figure 1.1**). These systems are passive in that they perform with no outside intervention or mechanical support. Passive fire protection cannot warn occupants of the dangers of an unwanted fire or suppress a growing fire. Meanwhile, active fire protection systems require some type of outside intervention or mechanical support, such as electricity or a water supply.

Active fire protection systems are fire protection, detection, and suppression systems that activate and operate automatically or manually and alert occupants upon the presence of smoke or fire. These systems, when activated, are designed to change the course and outcome of a fire in a building (**Figure 1.2**). These systems work to the benefit of the building owner, its occupants, and the responding fire department personnel. Active systems are the focus of this book.

Active fire protection systems can be divided into two broad categories: fire detection and fire suppression. **Fire detection systems** alert occupants to hazardous conditions, smoke, or changing fire conditions by automatic and/ or manual fire alarm equipment. The early detection of a fire and the signaling of an appropriate alarm remains the most significant factor in preventing large losses due to fire. **Fire suppression systems** are those that control and contain hazardous conditions until manual suppression can be achieved. Many of these systems are water-based. Working together, fire detection and suppression systems save lives and decrease property loss by notification and early mitigation and control of the fire.

Fire Detection System — System of detection devices, wiring, and supervisory equipment used for detecting fire or products of combustion and then signaling that these elements are present.

Fire Suppression System — System designed to act directly upon the hazard to mitigate or eliminate it, not simply to detect its presence and/or initiate an alarm.

Figure 1.1 Fire doors undergoing fire resistance testing.

Figure 1.2 A sprinkler is an active fire protection device.

History of Fire Protection, Detection, and Suppression Systems

When looking at fire protection, the concept of detection and suppression systems is deeply rooted in history. The first fire alarm system was installed around 1851. This was a municipal alarm system that used the telegraph to notify fire departments of a fire. The first commercially successful automatic fire sprinkler system was patented in 1872. In 1902, George Andrew Darby, an electrical engineer, patented the first heat indicator, which sounded a fire alarm when activated. This early device paved the way for today's battery-powered smoke alarms, which Kenneth House and Randolph Smith introduced in 1969.

Many catastrophic fires throughout history have served to increase the promotion and use of fire detection and suppression systems. Many of these events prompted changes in codes and regulations concerning the design, installation, and maintenance of fire protection systems as well as more restrictive requirements on assembly occupancies. These devastating events have become benchmarks in the history of fire protection, detection, and suppression systems and provide much evidence as to the lifesaving benefits of these systems. Notable fires include the following:

- **Iroquois Theater Fire, Chicago, Illinois, 1903** — This fire claimed the lives of 602 people. The scenery in the playhouse was made of canvas and painted in highly flammable oil paints. There were no automatic sprinklers for the stage, and the stage's fire curtain did not close properly. No emergency lighting was present, the smoke and heat vents for the stage were not functional, and many exit doors were either locked or did not open in the outward direction.

- **Triangle Shirtwaist Factory, New York City, 1911** — This company was located in a high-rise building, occupying the 8th, 9th, and 10th floors. An employee tossed a cigarette into a bin containing scrap material, causing it to ignite immediately. While employees on the 8th and 10th floors were able to escape due to early discovery and notification, the 250 employees on the 9th floor were not as fortunate. They were not informed of the fire and the dangerous situation it presented until the 8th floor was well involved but not the entire structure. The fire occurred at 4:45 p.m., around the time when the business was closing for the day. As was the custom to deter theft, security guards had already locked one of the two exit doors on the 9th floor. Some individuals were able to escape from the unlocked exit, but it was quickly made inaccessible by the fire. In addition, doors not locked were blocked, and the rear fire escape failed shortly after its use. The fire killed 145 people.

- **Cocoanut Grove Nightclub, Boston, Massachusetts, 1942** — A fire in this nightclub claimed the lives of 492 individuals. The NFPA Journal, May/June 2011 indicates that one account of how the fire started was that a busboy lit a match in a downstairs lounge while fumbling to find the socket for a small lightbulb tucked inside a fake palm tree. Seconds later, the palm tree was in flames and ignited the lounge's cloth ceiling. Within minutes, a huge fireball was roaring through the entire club. However, the cause of the Cocoanut Grove fire was deemed to be of "unknown origin."

  The nightclub was filled to twice its capacity, having approximately 1,000 occupants at the time the fire erupted. As with the Iroquois Theater fire, there were no automatic sprinklers, the exit doors did not swing in the proper direction, many doors and windows were sealed shut, and the primary exit was a revolving door.

- **Winecoff Hotel, Atlanta, Georgia, 1946** — This hotel was the scene of the deadliest hotel fire in United States history, causing 119 fatalities. The building had only one exit stairway, which became impassable early in the fire's development. Doors had been propped open, and the building did not have a fire suppression system. The building was also not equipped with a fire alarm system, so there was no way to notify the hotel's occupants when the fire started in the early morning hours.

- **Katie Jane Memorial Home, Warrenton, Missouri, 1957** — This event was one of the deadliest nursing home fires in American history. Of the 149 elderly residents, 71 perished in the fire and one died later at the hospital. The fire was believed to have started due to faulty wiring in the ceiling or in a wall. The facility did not have a sprinkler system, nor was there an alarm system, manual or automatic. The only means of escape for occupants were unenclosed interior stairways that emptied into the main building, and the facility lacked outside fire stairs or slides.

- **Our Lady of Angels School, Chicago, Illinois, 1958** — The deadliest school fire in United States history started in a basement trash can in the stairwell, and the structure's wooden staircase was quickly engulfed in flames. The location of the fire blocked the escape route for the occupants of the second floor. The exit corridors had combustible walls and ceilings, and again there were no automatic fire sprinklers. In addition, the school did not have an automatic fire alarm system, the stairwell was not enclosed, and a delay in notification of the fire department occurred. The fire resulted in the deaths of 92 children and three nuns.

- **Beverly Hills Supper Club, Southgate, Kentucky, 1977** — This incident was another nightclub fire that impacted codes and regulations concerning fire protection systems. It was Memorial Day weekend in 1977, and the building was packed with over 3,000 patrons. There was inadequate egress and exit identification and no fire suppression or alarm systems were present. The fire started in the Zebra Room, an unoccupied room, and was believed to be electrical in nature. The fire burned for some time before being discovered and quickly spread to other parts of the building. The fire took the lives of 165 individuals, many of whom were enjoying a show in the Cabaret Room at the opposite end of the building. Employees attempted to fight the fire prior to notifying the fire department or the building's occupants.

- **MGM Grand Hotel, Las Vegas, Nevada, 1980** — A fire at this hotel caused 87 fatalities and almost 700 injuries. While the hotel portion of the building was protected by a sprinkler system, the casino area did not have sprinklers. The hotel contained many unprotected vertical shafts and had openings that allowed smoke to enter and fill the exit stairwells. Once the occupants entered the stairwells, they were unable to return to their floor or room due to automatic locking doors.

- **Station Nightclub, West Warwick, Rhode Island, 2003** — One of the most recent notable fires in the United States took the lives of 100 patrons of this nightclub. Pyrotechnics used during a band performance ignited soundproofing foam at the back of the stage. Along with the nightclub being without automatic sprinklers, individuals panicked and failed to use secondary exit routes. People rushed for the front door, which became jammed, and most of the victims died trying to exit at the front of the club.

The necessity of fire protection, detection, and suppression systems is made obvious by the large loss of life in these events as well as the need for early notification of a building's occupants and the fire department. As a result of these and other similar events, codes and regulations concerning both active and passive fire protection systems have been adopted, modified, or more strictly enforced.

In addition to these historical benchmarks, two early reports have influenced the requirements for fire protection, detection, and suppression systems. These reports were prepared by groups that came together to address the unacceptable fire problem in America, and both reports promoted the use of fire protection devices for life safety and loss prevention during fires.

President Harry Truman called for and participated in the 1947 Fire Prevention Conference. This group was brought together to review the fire problem in the United States and to identify deficiencies in America's fire service. The purpose of the group was to determine ways to increase both the level of awareness about the fire problem and the "work of fire safety in every community." Regarding fire protection, detection, and suppression systems, the conference's report stated that all municipal fire prevention ordinances should be combined into one Fire Prevention Code. The report recommended better building design, the use of technology for prevention, and an increased focus on prevention education. Specifically, the report recommended the widespread use of fire extinguishing equipment, hoses, standpipes, automatic sprinklers, and alarm systems in buildings.

The current concept of the "Three E's of Prevention" — education, engineering, and enforcement — came as a result of this conference report. Individuals involved in fire prevention practices realized that all three components must be implemented in order to reduce the number of fire deaths and related injuries. As a result, fire prevention programs achieved more successful outcomes **(Figure 1.3)**.

President Richard Nixon convened the National Commission on Fire Prevention and Control in the 1970s. This commission produced the famous report entitled *America Burning*, which has had a significant impact on the fire service, fire prevention, and fire protection in the United States. Overall recommendations included the improvement of fire protection features of buildings. Specifically, this report identified the

Figure 1.3 Students in a kindergarten class learn about fire prevention.

need for "automatic fire extinguishing systems in every high-rise building and every low-rise building where people congregate." The report recommended that economic incentives be provided for built-in fire protection.

This report resulted in the passage of the Federal Fire Prevention and Control Act of 1974. The impact of this publication has been enormous in the areas of fire prevention, fire protection, fire service training, and public education. Since that time, the requirements for fire suppression and detection systems in commercial and multifamily residential structures have been incorporated and expanded. Additionally, many states and local jurisdictions have codes and regulations addressing requirements for such systems in one- and two-family dwellings.

Advocating for Increased Fire Prevention — Online Resources

Vision 20/20 offers free materials and guidance in its online Advocating for Fire Prevention toolkit. These resources were developed specifically to help the fire service gain advocates who can help increase needed resources and encourage broader participation in their fire prevention efforts.

One of the most important and challenging aspects of community risk reduction is the ability to convey the value of investing in public safety and fire prevention to a broad audience. With increased demand for services and with fewer resources, decision-makers and the public must understand the value of prevention and give their support for programs that can prevent and reduce losses from fire and other risks.

Content in the toolkit is organized according to these basic steps: 1) demonstrate need, 2) demonstrate results, and 3) develop relationships. The many resources within the toolkit can help make the case for investing in prevention, including:

● Links to reliable national data sources to describe risk in numbers

● Sample templates for using local data to focus attention on the need and the benefits of prevention

● Case studies from programs that document how prevention has saved lives, property, and the community

● Tools and guidelines for making effective presentations and working with the media

● Evidence of how partnerships can increase reach and save resources

● Two short videos for download:

— *Advocating for Prevention* illustrates the need for advocacy and encourages those individuals within the fire service to get started with an advocacy strategy.

— *Prevention Saves* demonstrates to the community how investment in prevention saves the things most important to it.

The Advocacy toolkit, produced and updated with AFG funding, is an online resource available at www.strategicfire.org.

Fire Protection, Detection, and Suppression Systems

A variety of fire protection, detection, and suppression systems exist. These systems can be divided into four basic categories:

- **Automatic fire protection, detection, and alarm systems** — Are activated by smoke, high temperatures, radiant heat, or flame signatures, and alert the occupants in the building and/or the fire department to the fire. These systems do not slow the growth of the fire or reduce the amount of smoke produced.

- **Automatic fire suppression systems** — Work to suppress the fire hazard by applying a fire-suppressing medium to the fire. Human intervention is not needed for their effectiveness. This process reduces the hazard to the occupants, to the building and its contents, and to responding firefighters.

- **Manual fire alarm systems** — Require activation by individuals upon discovery of a fire (**Figure 1.4**). Devices such as manual pull stations are common components of manual systems.

- **Manual fire suppression systems** — Include standpipes (both vertical and horizontal), fire hoses, and fire extinguishers (**Figure 1.5**). These mechanisms require an individual to operate them in order to perform their function; they are not automatic.

Figure 1.4 This manual pull station requires activation by an individual upon discovery of fire.

Figure 1.5 Class III standpipes are an example of a manual fire suppression system.

Standards and Codes for Detection and Suppression Systems

Code — A collection of rules and regulations enacted by a legislative body to become law in a particular jurisdiction.

As previously indicated, standards and **codes** have traditionally been developed and adopted in response to historical tragic events. Standards and codes are used to establish minimum requirements for the design and construction of buildings, structures, and facilities. They are also used to determine the type and extent of detection and suppression systems needed.

Standards and codes establish the minimum level of safety that should be present in a structure. The authority having jurisdiction (AHJ) must adopt this minimum, which code officials enforce.

Organizations that specialize in the development of standards and codes often create them. In order to be enforceable by law, the AHJ must adopt standards and codes.

Adoption of Standards and Codes

Transcription — Method by which an AHJ adopts a code in whole to become a new regulation.

Building and fire codes may be adopted in two ways. The first way is by **transcription**, where the entire code is copied into a regulation. The other method is by reference, where the regulation states that the referenced code is legally enforceable as part of the fire and life safety regulations. Codes can be adopted in whole or in part and may be amended by the jurisdiction through the adoption process if allowed by governing authorities.

Consensus Standard — Rules, principles, or measures that are established though agreement of members of the standards-setting organization.

Standards attempt to obtain consistency in design, practice, and materials and can provide a guide of practices and designs that have proven to be successful. Many standards are **consensus standards**, meaning that a group of experts has developed and agreed upon the standard before it is adopted.

Before codes were developed, a wide variety of interpretations and even confusion for requirements existed. Model codes were developed to provide agreed-upon requirements for areas, such as fire and life safety or electrical equipment. These codes are similar to consensus standards in that a group agrees upon the code before it is adopted into the model.

Standards-and Codes-Developing Organizations

A variety of organizations publish consensus standards and model codes. The National Fire Protection Association® (NFPA®) is the most familiar organization. The NFPA® publishes the majority of the consensus standards used in the United States and Canada. Since the 1880s, the NFPA® has published a variety of standards and other documents dealing with fire prevention and building safety. Many of these standards address the various fire detection and suppression systems that are required for specific occupancies or buildings.

The International Code Council (ICC) is a membership association dedicated to building safety and fire prevention. It was established in 1994 as a nonprofit organization that is dedicated to developing a single set of comprehensive national model construction codes. The ICC develops the codes used to construct residential and commercial buildings. The council's founders include the Building Officials and Code Administrators International, Inc. (BOCA), International Conference of Building Officials (ICBO), and Southern Building Code Congress International, Inc. (SBCCI). These three organizations had previously developed separate sets of model codes that were used throughout the United States.

The Canadian Commission on Building and Fire Codes (CCBFC) develops and maintains Canada's model codes. Canada has six model construction and fire codes, including the National Fire Code of Canada (NFC) and the National Building Code of Canada (NBC). These comprehensive building and fire safety regulations are used throughout Canada and address construction and fire protection systems requirements for buildings and occupancy classifications.

ASTM International (originally known as the American Society for Testing and Materials) is one of the largest standard developing organizations and is considered to be a trusted source for technical standards for materials, products, systems, and services. ASTM International develops testing processes that other testing organizations use in the development of safety products.

Underwriters Laboratories Inc. (UL) is an independent, not-for-profit product certification organization that has been testing products and writing standards for safety since 1894. UL annually evaluates thousands of products, components, materials, and systems **(Figure 1.6)**. UL has developed over 800 standards for safety. Some of these standards relate directly to fire and life safety, such as smoke detectors for fire alarm systems and dry chemical fire extinguishers.

FM Global (formerly Factory Mutual) is a property insurance company that works to provide property improvements for business owners to prevent risk. The organization works in partnership with businesses and industries to reduce property damage through state-of-the-art property-loss prevention research and engineering and comprehensive insurance products. FM Global provides product certification and testing services through an approval process and specification testing.

The U.S. Occupational Safety and Health Administration (OSHA) recognizes nationally recognized testing laboratories (NRTLs) as having the capability to provide product safety testing and certification services to the manufacturers of a wide range of products. The testing and certifications are based on

Figure 1.6 The results of a flame test conducted on cable.

widely accepted product safety standards and are often issued by the American National Standards Institute (ANSI). These testing agencies must meet the certification requirements of OSHA's Directorate of Technical Support and Emergency Management. Both UL and FM Global are certified as NRTLs through OSHA. Other smaller testing laboratories include Southwest Research Institute (SwRI), Wyle Laboratories (WL), and the Canadian Standards Association (CSA). OSHA provides a complete list of these agencies on its website. ANSI coordinates the development and use of voluntary consensus standards in the United States. ANSI is also involved in accrediting programs that assess conformance to standards in businesses, including the International Organization for Standardization (ISO) 9000 and ISO 14000 management systems. ANSI standards may be cross-referenced between NFPA® and OSHA documents.

Jurisdictions base their building and fire codes from this wide array of sources. These codes designate the degree or level of protection required in a building based upon its occupancy, construction type, and other designated variables.

Benefits of Fire Protection, Detection, and Suppression Systems

The influence of fire protection, detection, and suppression systems on life safety cannot be overstated. The positive effects for both building occupants and firefighters are readily apparent by an examination of historical data. In each of the fires discussed previously, the investigations revealed the lack of fire alarm systems and fire suppression systems as a major contributing factor in the deadly results.

According to estimates by the NFPA® and the United States Fire Administration (USFA), U.S. home usage of smoke alarms rose from less than 10 percent in 1975 to 94 percent in 2000. This data was drawn from USFA Topical Fire Research Series Vol 1, Issue 15, Smoke Alarm Performance in Residential Structure Fires.

At the same time, the number of home fire deaths was cut nearly in half. The relationship between improved use of smoke alarms and lower death rates due to fire is not a coincidence. Because of this drastic reduction in fire deaths, the home smoke alarm is widely credited as one of the greatest successes in fire safety in the last part of the 20th century **(Figure 1.7)**. The NFPA® states that a properly installed and maintained automatic sprinkler system will reduce the average property loss from fire by one-half to two-thirds. According to the USFA, property losses are 85 percent less in buildings that are protected with fire sprinklers compared to those buildings without sprinklers. The combination of automatic sprinklers and early warning systems in all buildings and residences could reduce injuries, loss of life, and property damage by at least 50 percent. The USFA has promoted research, development, testing, and demonstrations of residential fire sprinkler systems for more than 30 years. USFA research regarding residential fire sprinkler systems shows that these systems can do the following:

- Save the lives of building occupants
- Save the lives of firefighters called to respond to a home fire

Figure 1.7 Fire department personnel often install smoke alarms for free in their communities.

- Significantly offset the risk of premature building collapse posed to firefighters by lightweight construction components when they are involved by fire

- Substantially reduce property loss caused by a fire

The Fire and Life Safety Section of the International Association of Fire Chiefs (IAFC) has stated priorities for the promotion of fire and life safety. These priorities include reducing death from structure fires to zero and limiting property damage from structures to the area of origin as well as reducing firefighter fireground deaths to zero. In pursuit of these priorities, the IAFC advocates complete automatic fire sprinkler protection in all new occupied construction, including one- and two-family dwellings. The association also advocates fire sprinkler retrofits of existing high-rises, institutional, and other high-risk/high-consequence occupancies. In addition, the IAFC recommends educating all firefighters on the benefits of fire sprinkler protection in reducing firefighter deaths.

In addition to protecting civilians and their properties, there are also benefits to emergency responders. In its work to reduce the number of firefighter deaths and injuries, the National Fallen Firefighters Foundation, through its Everyone Goes Home® Firefighter Life Safety Initiative Program, states that "advocacy must be strengthened for the enforcement of codes and the installation of home fire sprinklers." This program lists 16 initiatives that can be incorporated into a fire and emergency services organization to help meet the program's objectives.

Chapter Summary

Fire protection, detection, and suppression systems are of vital importance in the goal of fire prevention. While these systems may be unable to prevent a fire from occurring, they work effectively and efficiently to prevent loss of life or large property loss. History demonstrates the seriousness of the fire problem and the tragic toll it can take on a community or fire department. These systems allow fire professionals to achieve their goal of life safety and property protection.

Review Questions

1. What is the difference between fire detection and fire suppression systems?

2. What are some events that have become benchmarks in the history of fire detection and fire suppression systems?

3. What are the four basic categories of fire protection, detection, and suppression systems?

4. What are the two ways that building and fire codes can be adopted?

5. How are automatic sprinkler systems beneficial to life safety and loss prevention?

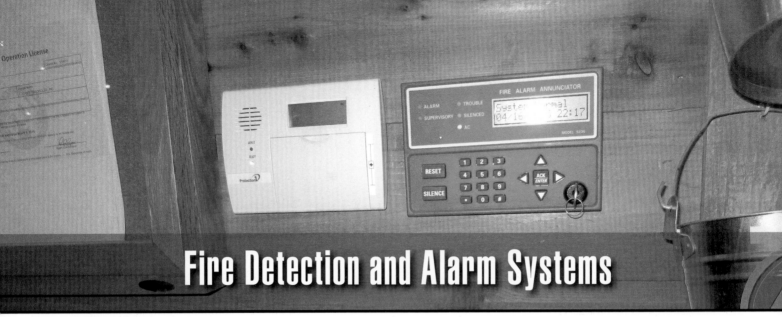

Fire Detection and Alarm Systems

Courtesy of Rich Mahaney.

Chapter Contents

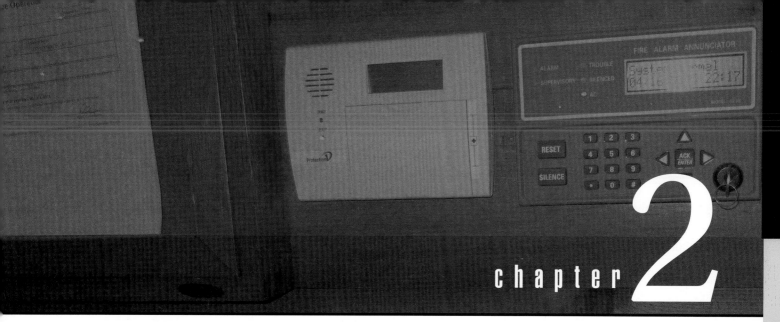

Key Terms

FESHE Outcomes

Fire and Emergency Services Higher Education (FESHE) Outcomes:
Fire Protection Systems

7. Explain the basic components of a fire alarm system.

8. Identify the different types of detectors and explain how they detect fire.

Fire Detection and Alarm Systems

Learning Objectives

After reading this chapter, students will be able to:

1. Describe the basic components of fire detection and alarm systems.

2. Identify major types of fire detection and alarm systems.

3. Identify the three methods of transmitting fire alarm signals.

4. Describe the methods of emergency response notification and alarm identification within the protected premise.

5. Explain the operation of manual fire alarm boxes.

6. Describe types of automatic alarm-initiating devices.

7. Describe methods for testing and inspecting fire detection and alarm systems.

8. Summarize the requirements and timetables for the inspection and testing of various types of systems.

Chapter 2
Fire Detection and Alarm Systems

Courtesy of Rich Mahaney.

Case History

On July 10, 2015, at 4:18 a.m., an alarm monitoring company received a signal from a hotel's automatic fire alarm system. The hotel is a three-story building comprised of three separate buildings without a fire sprinkler system. At the time of the fire, the hotel was at maximum capacity. The alarm company received a signal for a smoke detector activation on the first floor of the hotel. A few moments later, the alarm company received a second signal for an additional smoke detector activation on the first floor. Very soon after the second signal, the alarm company received a third signal from a pull station activation on the second floor of the hotel.

Upon the fire crews arrival, firefighters discovered a large fire rapidly progressing up the exterior of the building. Fire crews made an attack on the fire and quickly extinguished the flames. The outcome might have been vastly different without the presence of a fire alarm system within this building. The fire alarm system greatly contributed to the detection of the fire and notification and evacuation of the building's occupants.

This chapter provides information on the basic components and types of fire protection, detection, and alarm systems, both manual and automatic. In addition, the inspection and testing of these systems will be described as well as the importance of record keeping and timetables.

Fire Detection and Alarm System Components

Modern detection and signaling systems vary in complexity from those that are simple to those that incorporate advanced detection and signaling equipment. Such systems are typically designed and installed by qualified individuals as determined by the authority having jurisdiction (AHJ). The design, installation, and approval of a fire detection and alarm system may also require acceptance testing by regulatory agencies before new buildings can be occupied or the system can be placed in service.

A nationally recognized testing laboratory, such as Underwriters Laboratories Inc. (UL), FM Global, or Intertek (ETL Listed Mark), should test the components of a system to ensure operational reliability **(Figure 2.1, p. 28)**.

Figure 2.1 UL personnel preparing to test fire detection and alarm system wiring under fire conditions.

In addition, the installation of the system should conform to the applicable provisions of NFPA® 70, *National Electrical Code®*; NFPA® 72, *National Fire Alarm and Signaling Code*, and local codes and ordinances.

Several basic components, such as fire alarm control units (FACUs), primary and secondary power supplies, initiating devices, and notification appliances, are found in various fire detection and alarm systems. The following sections explain in more detail the components of these systems.

Fire Alarm Control Units (FACUs)

The *fire alarm control unit (FACU)*, formerly called the fire alarm control panel (FACP), contains the electronics that supervise and monitor the integrity of the wiring and components of the fire alarm system. The FACU basically serves as the brain for the alarm system **(Figure 2.2)**. It receives signals from alarm initiating devices, processes the signals, and produces output signals that activate audible and visual appliances and transmit signals to an off-site monitoring station when provided. Power and fire alarm circuits are connected directly into this panel. In addition, the remote auxiliary fire control units and notification appliance

Fire Alarm Control Unit (FACU) — The main fire alarm system component that monitors equipment and circuits, receives input signals from initiating devices, activates notification appliances, and transmits signals off-site. Formerly called the *fire alarm control panel (FACP)*.

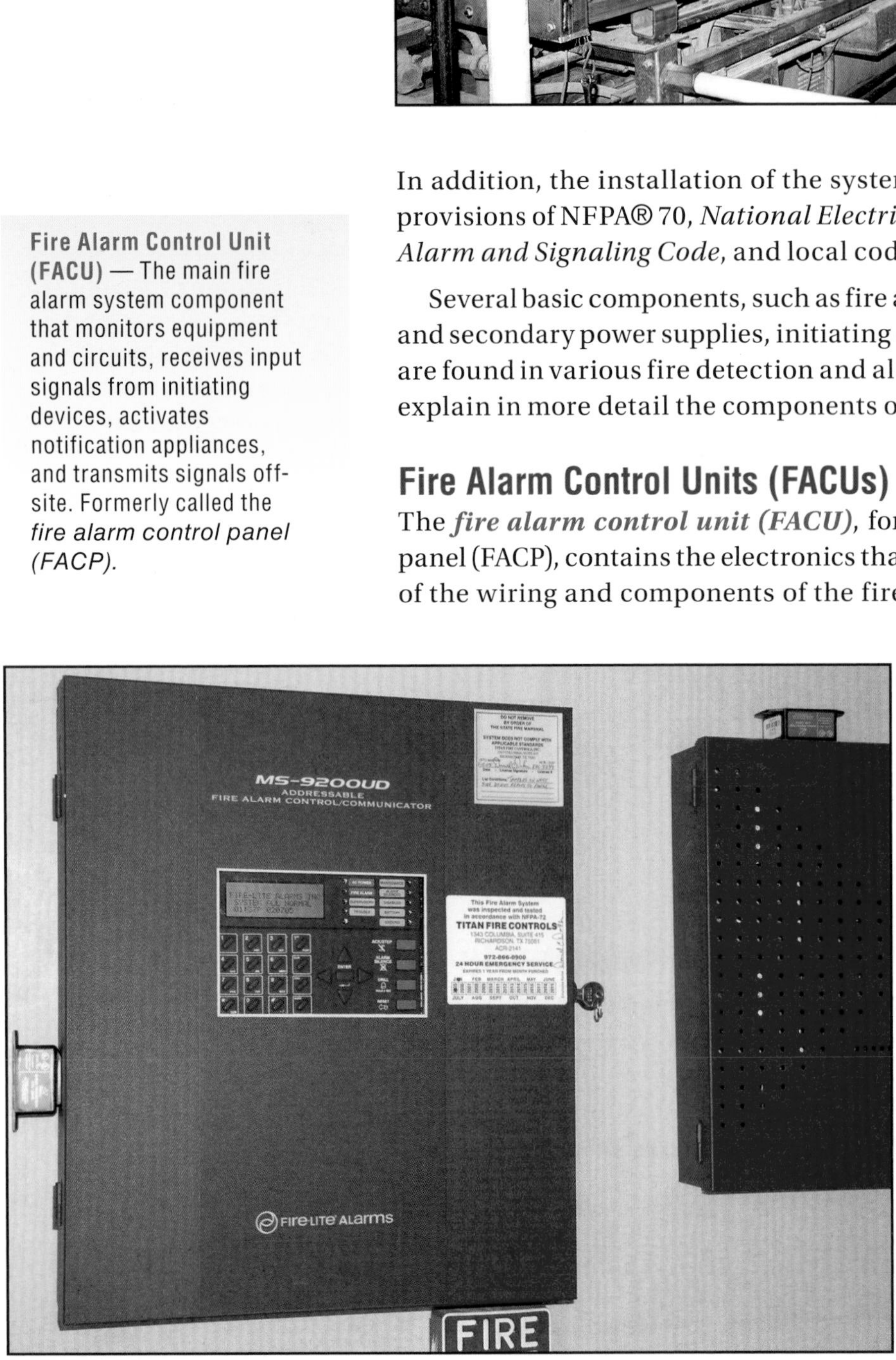

Figure 2.2 The fire alarm control unit (FACU) contains the electronics that supervise and monitor the integrity of the wiring and components of the fire alarm system. *Courtesy of Ron Moore, McKinney (TX) Fire Department.*

panels are considered to be part of the fire alarm system and are connected and controlled.

Controls for the system are located in the FACU. The FACU can also perform other functions, such as provide two-way firefighter communication, remote annunciator integration, control of elevators, HVAC, fire doors, dampers, locks, or other fire protection features. Public address messages and mass notifications alerts can also be accomplished through prerecorded evacuation messages or independent voice communications from the FACU.

Primary Power Supply

The primary electrical power supply usually is obtained from the building's main power connection to the local utility provider **(Figure 2.3)**. In rare instances where electrical service is unavailable or unreliable, an engine driven generator can provide the primary power supply. If such a generator is used, either a trained operator must be on duty 24-hours a day or the system must contain multiple engine-driven generators. One of these generators must always be set for automatic starting. The FACU must supervise the primary power supply and activate a trouble signal if the power is interrupted.

Secondary Power Supply

A secondary power supply must be provided for all fire alarm systems. This requirement is done to ensure that the system will be operational even if the main power supply fails. The secondary power supply must be capable of providing normal, standby conditions (nonalarm) capacity and power to fully operate an alarm condition for the required duration. The time period requirements for secondary power operation capabilities vary and can be found in NFPA® 72. Secondary power sources can consist of batteries with chargers, engine-driven generators with a storage battery, or multiple engine-driven generators, of which one must be set for automatic starting **(Figure 2.4)**.

Initiating Devices

Initiating devices are the manual and automatic devices that are activated or sense the presence of the products of combustion or other hazardous conditions and then send a signal to the FACU. The initiating devices may be either connected to the FACU by the hard-wire system or radio-controlled over a special frequency. Initiating devices include but are not limited to:

- Manual pull stations **(Figure 2.5, p. 30)**
- Heat detectors

Figure 2.3 Power for the system is obtained from the local utility provider.

Figure 2.4 Generators are a common secondary power source. *Courtesy of Ron Moore, McKinney (TX) Fire Department.*

Figure 2.5 A manual pull station is one type of initiating device.

- Smoke detectors
- Flame detectors
- Waterflow devices
- Tamper switches
- Combination detectors
- Other specialty suppression systems

NOTE: This chapter will explain initiating devices in more detail.

Notification Appliances

Audible notification signaling appliances are the most common types of alarm-signaling systems used for signaling a fire alarm in a structure. Once an alarm initiating device is activated, it sends a signal to the FACU, which then processes the signal and initiates actions. The primary action initiated is usually local notification, which can take the form of:

- Bells
- Buzzers
- Horns
- Speakers
- Strobe lights
- Other warning appliances

Depending on the system's design, the local alarm may either activate a single notification appliance, notification appliances within a specific zone, designated floor(s), or the entire facility. Notification appliances fall under the following categories **(Figure 2.6)**:

- **Audible** — Approved sounding devices, such as horns, bells, or speakers, that indicate a fire or emergency condition

- **Visual** — Approved lighting devices, such as strobes or flashing lights, that indicate a fire or emergency condition

- **Textual** — Visual text or symbols indicating a fire or emergency condition

- **Tactile** — Indication of a fire or emergency condition through sense of touch or vibration **(Figure 2.7)**

Off-Site Reporting Methods

Several different methods are used to send signal transmissions from the facility fire alarm system to a public or private reporting station. These methods include:

- **Digital alarm communicator transmitter (DACT)** — Provides transmission through phone lines to a monitoring station.

- **Digital alarm radio system (DARS)** — Sends a radio signal from a digital radio transmitter located at the protected premises to a monitoring station.

- **Cellular** — Uses cellular phone technology to transmit a signal from the protected premises to a monitoring station.

- **Internet protocol** — Transmits a signal from the protected premises to a monitoring station through an approved Internet connection.

Figure 2.6 This appliance incorporates both audible and visual signals.

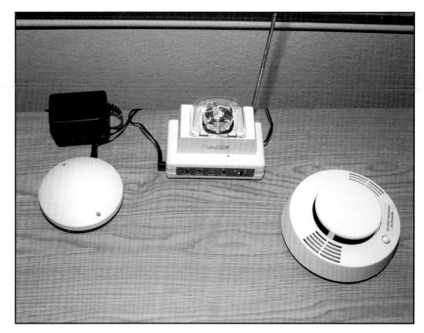

Figure 2.7 Bed shakers are integrated with the detection device in the occupancy and alert those who are deaf or hard-of-hearing to an emergency.

- **City tie/polarity reversal** — Uses a hardwired system that communicates a signal from the protected premises directly to the emergency services telecommunications center or other approved location.

Additional Alarm System Functions

Building codes have special requirements for some types of occupancies in the event of a possible fire condition. In these cases, the fire detection and alarm system can be designed to initiate the following actions:

- Turn off the heating, ventilating, and air-conditioning (HVAC) system.

Figure 2.8 Smoke dampers can be closed remotely by the system.

- Close **smoke dampers** and/or fire doors **(Figure 2.8)**.

- Pressurize stairwells and/or operate smoke control systems for evacuation purposes.

- Unlock doors along the path of egress.

- Provide elevator recall to the designated floor and prevent normal operations.

- Operate heat and smoke vents.

- Activate special fire suppression systems, such as preaction and deluge sprinkler systems or a variety of special-agent fire extinguishing systems.

Smoke Damper — Device that restricts the flow of smoke through an air-handling system; usually activated by the building's fire alarm signaling system.

Types of Fire Detection and Alarm Systems

Fire detection and alarm systems are designed to receive certain types of signals from devices and perform an action based upon the type of signal received. Some signals may indicate a fire condition, while others may indicate that a device on the system needs to be serviced. The FACU should be programmed to respond to different signal types in an appropriate manner.

For many years, fire alarm-initiating devices were neither analog nor addressable. Since they were nonanalog, there were only two operating states: normal and alarm. Because they were not addressable, conventional initiating devices were unable to provide a specific location to the FACU. Today, modern initiating devices may have analog capabilities, meaning they can monitor the conditions in a protected space. In addition, addressable initiating devices are available, which will communicate the specific location of the device that is activated to the FACU.

Fire detection and alarm systems are designed to produce three types of specialty signals, depending on the type and nature of the alarm they are reporting.

- An **alarm signal** is a warning of a fire emergency or dangerous condition that demands immediate attention.

- A **supervisory signal** indicates an off-normal condition of the complete fire protection system. Supervisory signals also include a returned-to-normal signal, meaning that the condition has been resolved. These signals are used to monitor the integrity of the fire protection features of the system.

- A **trouble signal** indicates a problem with the fire alarm system or the system's power supply. Each signal must be audibly and visually displayed at the FACU in a distinct manner that dif-ferentiates one type of signal from another.

NOTE: A trouble signal indicates a problem with the fire detection and alarm system. A supervisory alarm indicates a problem with an accessory of the fire alarm system.

Alarm Signal — Signal given by a fire detection and alarm system when there is a fire condition detected.

Supervisory Signal — Signal given by a fire detection and alarm system when a monitored condition in the system is off-normal.

Trouble Signal — Signal given by a fire detection and alarm system when a power failure or other system malfunction occurs.

Household Fire Alarm Systems

Initially evaluated by the Los Angeles Fire Department and NFPA® in the early 1960s, household fire warning equipment was adopted as an NFPA® standard in 1974. Early studies indicated a 35 percent increase in lives saved and an 80 percent reduction in property damage from fires when residential smoke alarms were installed; the understanding was that approximately 71 percent of fires occurred in residential occupancies, and residential fires accounted for nearly 78 percent of all fire deaths and 57 percent of the total dollar loss. By 1974, most jurisdictions began requiring smoke alarms within new residential dwellings with the advent of affordable battery operated and in-line smoke alarms and changes in public acceptance to curb residential fire losses. A 2006 White Paper written by the Public/Private Fire Safety Council indicates 96 percent of all U.S. homes have smoke alarms installed. The 4 percent of homes remaining without smoke alarms are responsible for 39 percent of all reported home fires and nearly 50 percent of all home fire deaths.

Smoke Alarms vs. Smoke Detectors

The terms *smoke alarm* and *smoke detector* are often used interchangeably. While this is common practice, it is technically incorrect. Smoke alarms are the devices typically installed in residential occupancies. These devices combine a smoke detector with a local notification appliance. When activated, smoke alarms emit an audible alarm to notify occupants of the presence of smoke.

Smoke detectors differ from smoke alarms in that they do not include a local notification appliance. When activated, smoke detectors send a signal to an FACU or a similar device. The FACU then initiates the alarm to notify occupants.

Multisensor vs. Multicriteria Detectors

Both multisensor and multicriteria detectors employ multiple sensors within one device that respond to physical stimulus such as heat, smoke, carbon monoxide $(CO)_2$, or other fire gases. Multisensor detectors contain multiple sensors that can transmit multiple alarm signals to the FACU. Multicriteria detectors employ more than one sensor and are only capable of generating one alarm signal, based on the sensors in the device responding independently or in combination to a fire signature. Both types of detectors rely on a mathematic evaluation of the output signals from the sensors to determine when an alarm signal shall be transmitted.

Single- vs. Multi-Station Smoke Alarms

Single-station smoke alarms are stand-alone devices. When smoke is detected at the location of one device, only that device sounds and alerts occupants to the presence of smoke. Multistation smoke alarms are interconnected devices. If one device detects the presence of smoke, all devices will sound and alert occupants in all areas within a dwelling unit. These are the two most common arrangements of smoke alarms in household fire alarm systems; however, these arrangements may also be found on other types of fire alarm systems in residential occupancies. Initially required to be installed outside of sleeping rooms, most jurisdictions now require smoke alarms to be installed within sleeping rooms and on each level of the dwelling unit. Refer to NFPA® 72 for specific installation requirements.

Carbon Monoxide Detectors and Alarms

Carbon monoxide is a colorless, tasteless, and odorless gas that is often referred to as the "silent killer." According to the Center for Disease Control and Prevention, more than 400 Americans die from unintentional CO poisoning, while over 20,000 Americans visit emergency rooms and more than 4,000 are hospitalized due to CO poisoning. Increased CO poisoning awareness has brought about the increased installation of carbon monoxide detectors within households. A CO detector is a device that will detect the presence of a dangerously elevated level of carbon monoxide. Single station CO detectors with a sounding device are classified as a CO alarm. CO detectors that are system devices are referred to as *carbon monoxide detectors* by the listing laboratory.

Single station carbon monoxide alarms are typically used in residential locations. The alarm will sound when levels of CO elevate to levels that are dangerous to humans. With states recently adopting legislation and regulations (codes), the installation of CO detectors will increase in other occupancies such as schools. Local emergency responders must be aware of the increased use of these alarms and discuss their response tactics to such response requests.

As an optional level of protection, newer smoke alarms are equipped with a carbon monoxide sensor that detects and alarms when the CO level exceeds 70 parts per million (PPM). Carbon monoxide becomes a health hazard at levels above 150 ppm and is immediately dangerous to life and health (IDLH) at levels above 1,200 ppm.

Fire Alarm Signaling Systems

Fire alarm systems range from the simple to the complex. A simple system may only sound a local evacuation alarm. Whereas, a more complex system may sound a local alarm, activate building services, and notify fire and security agencies to respond. The type of system required depends upon the type of occupancy of the building and is affected by the following factors:

- Level of life safety hazard
- Structural features of the building
- Hazard level presented by the contents of the building
- Availability of fire suppression resources such as water supply, hydrants, and automatic sprinkler systems
- State and local code requirements

Individuals should be able to recognize each type of system and understand how each system operates. This recognition is important when performing inspections or conducting preincident planning. Several types of systems include the following:

- Protected premises (local)
- Supervising station alarm systems
- Proprietary
- Emergency communications systems

Both emergency communications systems and parallel telephone systems may be found in buildings with certain occupancies or building types. Mass notification systems are a special type of emergency communications systems that may be found as a part of a building's alarm system to provide specific and detailed instructions to a building's occupants. NFPA® 72 contains the requirements for all fire alarm and protective signaling systems and should be consulted for further information. The following section explains the major systems.

Protected Premises Systems (Local)

A **protected premises system** is designed to provide notification to building occupants only on the immediate premises. Where these systems are allowed, there are no provisions for automatic off-site reporting. The protected premise system can be activated by manual means, such as a pull station, or by auto-

Protected Premises System — Alarm system that alerts and notifies only occupants on the premises of the existence of a fire so that they can safely exit the building and call the fire department. If a response by a public safety agency (police or fire department) is required, an occupant hearing the alarm must notify the agency.

matic devices, such as smoke detectors. A protected premise system may also be capable of annunciating a supervisory or trouble condition to ensure that service interruptions do not go unnoticed. A protected premises system can be designed to activate the auxiliary services described later in this chapter.

Presignal Alarms

Presignal alarms are unique systems that may be employed in locations, such as hospitals, where greater assistance is needed to help occupants evacuate in a safe and orderly manner. The system initially responds with a presignal that alerts emergency personnel before the general occupancy is notified. This presignal is usually a discreet signal that is recognizable only by personnel who are familiar with the system. The presignal may be a recorded message over an intercom, a soft alarm signal, or a pager notification. The presignal provides emergency personnel with an opportunity to assist the general occupancy in evacuation. Depending on the policies of the occupancy and local code requirements, emergency personnel may elect to handle the incident without sounding a general alarm. Personnel may elect to sound the general alarm after investigating the problem, or the general alarm will sound automatically after a certain amount of time has passed and the fire alarm control unit has not been reset.

Conventional Alarm System. A conventional system is the simplest type of protected premises alarm system. When an alarm-initiating device, such as a smoke detector, sends a signal to the FACU, all of the alarm-signaling devices operate simultaneously **(Figure 2.9, p. 36)**. The signaling devices usually operate continuously until the FACU is reset. The FACU is incapable of identifying which initiating device triggered the alarm; therefore, building and fire department personnel must walk around the entire facility and visually check to see which device was activated. These systems are only practical in small occupancies with a limited number of rooms and initiating devices.

An FACU serves the premises as a local control unit. This system is found in occupancies that use the alarm signals for other purposes. In the past, schools sometimes used the same bells for class change as for fire alarms. The FACU enables the fire alarm to have a sound that is distinct from class bells, eliminating confusion as to which type of alarm is sounding. Modern codes do not allow systems such as these.

NOTE: However, older systems that use the same bells are still encountered.

Zoned Conventional Alarm System. Fire-alarm system annunciation enables emergency responders to identify the general location (zone) of alarm device activation. In this type of system, an annunciator panel, FACU, or a printout visibly indicates the building, floor, fire zone, or other area that coincides with the location of an operating alarm-initiating device **(Figure 2.10, p. 36)**.

Alarm-initiating devices in common areas are arranged in circuits or zones. Each zone has its own indicator light or display on the FACU. When an initiating device in a particular zone is triggered, the notification devices are activated, and the corresponding indicator is illuminated on the FACU. This signal gives responders a better idea of where the problem is located.

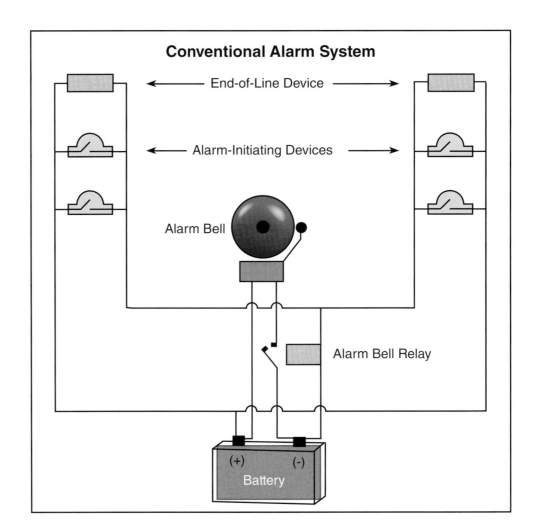

Conventional Alarm System

End-of-Line Device

Alarm-Initiating Devices

Alarm Bell

Alarm Bell Relay

(+)   (-)
Battery

Figure 2.9 This system cannot differentiate which alarm-initiating device caused the alarm.

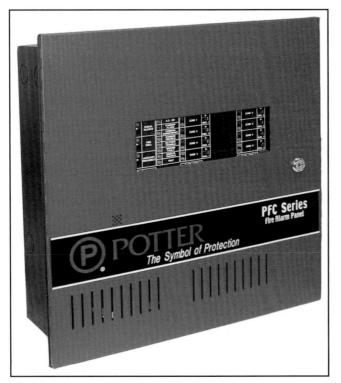

Figure 2.11 Addressable alarm systems identify the specific device that has been activated. *Courtesy of Potter Signal.*

Figure 2.10 A typical FACU for a zone system. *Courtesy of Potter Signal.*

A remote annunciator panel may be located some distance from the FACU, often in a location designated by the fire department. For example, such an installation may be found at the driveway approach to a large residential retirement complex. This type of annunciator panel usually has a map of the complex coordinated to the zone indicator light.

Addressable Alarm Systems. Addressable alarm systems display the location of each initiating device on the FACU and an annunciator panel if provided **(Figure 2.11)**. This connection ensures that firefighters or building personnel responding to the alarm can pinpoint the specific device that has been activated. Addressable systems reduce the amount of time that it takes to respond to emergency situations. These systems also allow repair personnel to quickly locate and correct malfunctions in the system.

Supervising Station Alarm Systems

A supervising station alarm system is continuously monitored at a remote location for the purpose of reporting a supervisory, trouble, or alarm signal to the appropriate authorities. Supervising station alarm systems are the predominant types of signal-monitoring systems used in the United States. Types of supervising station systems include auxiliary alarm systems, proprietary systems, central station systems, and remote receiving systems.

A **public emergency alarm reporting system** is connected to a municipal fire alarm system. Alarms are transmitted over this system to a public fire telecommunications center, where the appropriate response agencies are selected and dispatched to the alarm **(Figure 2.12)**. Two types of auxiliary fire alarm systems are the local energy system and the shunt-type auxiliary fire alarm system.

Public Emergency Alarm Reporting System — System that connects the protected property with the fire department alarm communications center by a municipal master fire alarm box or over a dedicated telephone line.

Figure 2.12 A public fire telecommunications center.

A local energy system has its own power source and does not depend on the supply source that powers the entire municipal fire alarm system. In these systems, initiating devices can be activated even when the power supply to the municipal system is interrupted. However, interruption may result in the alarm only being sounded locally and not being transmitted to the fire department telecommunications center. The ability to transmit alarms during power interruptions depends on the design of the municipal system.

A shunt system is electrically connected to an integral part of the municipal fire alarm system and depends on the municipal system's source of electric power. When a power failure occurs in this type of system, an alarm indication is sent to the fire department communications center. NFPA® 72 allows only manual pull stations and waterflow detection devices to be used on shunt systems. Fire detection devices are not permitted on a shunt system.

Proprietary Alarm System
— Fire alarm system owned and operated by the property owner.

Proprietary Systems

A **proprietary system** is used to protect large commercial and industrial buildings, high-rise structures, and groups of commonly owned facilities, such as a college campus or industrial complex in single or multiple locations **(Figure 2.13)**. Each building or area has its own system that is wired into a common receiving point that the facility owner operates. The receiving point must be in a separate structure or a part of a structure that is remote from any hazardous operations.

The receiving station of a proprietary system is continuously staffed by personnel who are trained in the system's operation and can take necessary actions upon an alarm activation. The operator should be able to automatically summon a fire department response through the system controls or by using the telephone. Many proprietary systems and receiving points are used to monitor security functions in addition to fire and life safety functions. Modern proprietary systems can be complex and have a wide range of capabilities, including coded-alarm and trouble-signal indications, building utility controls, elevator controls, and fire and smoke damper controls.

Parallel Telephone and Multiplexing Systems

Parallel Telephone Systems

A parallel telephone system consists of a dedicated circuit between each individual alarm box or protected property and the fire department telecommunications center. NFPA® 72 requires that these telephone systems are not used for any other purpose than to relay alarms. These systems are generally not found today due to the existence of private monitoring firms.

Multiplexing Systems

Multiplexing systems allow the transmission of multiple signals over a single line. This type of system allows the alarm-initiating devices to be identified individually, as in an addressable system, or in a group through the interaction of the fire alarm control panel with each independent device. Remote devices, such as relays, can be controlled over the same line to which initiating and indicating devices are connected. This connection greatly reduces the amount of circuit wiring needed for large applications.

The control panels for multiplexing systems can range from the simple and relatively inexpensive to the sophisticated and costly. Some multiplex systems have the added advantage of being able to test the performance of the devices, reducing manpower requirements for preventive maintenance.

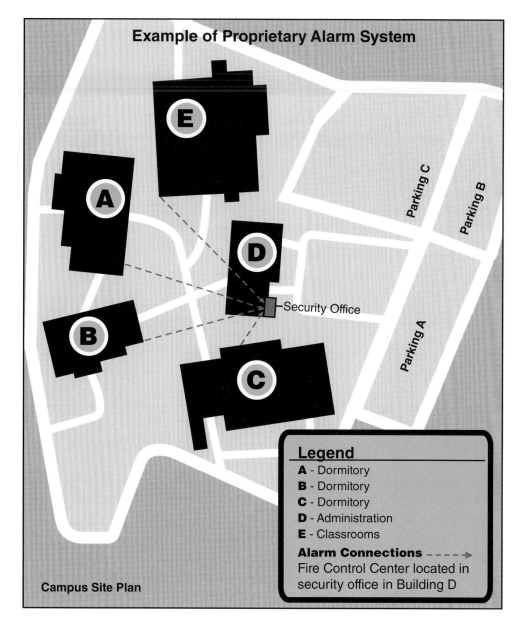

Example of Proprietary Alarm System

Parking C
Parking B
Parking A

Security Office

Legend
A - Dormitory
B - Dormitory
C - Dormitory
D - Administration
E - Classrooms

Alarm Connections - - - - →
Fire Control Center located in security office in Building D

Campus Site Plan

Figure 2.13 An example of a proprietary system being used to protect a college campus.

Central Station Systems. A **central station system** is a listed supervising station that monitors the status of protected premise alarm systems and provides inspection, testing, maintenance, and runner services through contracted services **(Figure 2.14, p. 40)**. The runner must be a qualified technician who can respond within two hours for an alarm or supervisory condition and within four hours for a trouble condition. All central station systems and communication methods should meet the requirements set forth in NFPA® 72. Central stations, when meeting the listing requirements, must be listed by an approved listing service.

Remote Receiving Systems. A listed supervising station that monitors the status of protected premise alarm systems through contracted services is called a *remote receiving station*. Remote receiving stations do not provide inspection, testing, maintenance, or runner services.

Depending on local requirements, the fire department may approve other organizations to monitor the remote system. In some small communities, the local emergency services telecommunications center monitors the system.

Central Station System — Alarm system that functions through a constantly attended location (central station) operated by an alarm company. Alarm signals from the protected property are received in the central station and are then retransmitted by trained personnel to the fire department alarm communications center.

Remote Receiving System — System in which alarm signals from the protected premises are transmitted over a leased telephone line to a remote receiving station with a 24-hour staff; usually the municipal fire department's alarm communications center.

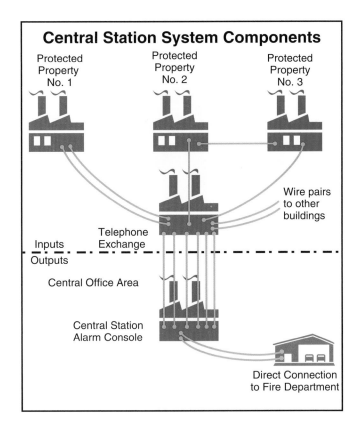

Central Station System Components

Protected Property No. 1

Protected Property No. 2

Protected Property No. 3

Wire pairs to other buildings

Telephone Exchange

Inputs

Outputs

Central Office Area

Central Station Alarm Console

Direct Connection to Fire Department

Figure 2.14 Components of a Central Station System.

This arrangement is particularly common in communities that have volunteer fire departments whose stations are not continuously staffed. In these cases, emergency services telecommunications personnel must be aware of the importance of these alarm signals and trained in the actions that must be taken upon alarm receipt.

Emergency Communications Systems

An emergency communications system is a supplementary system that may be provided in facilities in conjunction with detection and alarm signaling systems. The purpose of emergency communications systems is to provide a reliable communication system for occupants and firefighters. This system may either be a stand-alone system or it may be integrated directly into the overall fire detection and alarm-signaling system. System types include voice notification, two-way communication, and mass notification.

Voice Notification Systems. A one-way voice notification system warns building occupants that action is needed and tells them what action to take. This type is most commonly used in high-rise buildings, places of assembly, and educational occupancies. Occupants can be directed to move to areas of refuge in the building, leave the building, or stay where they are if they are in an unaffected area.

Two-Way Communication Systems. A two-way emergency communication system allows people, such as emergency personnel, at other locations in the building to communicate with the person at the fire command center using either intercom controls or special telephones **(Figure 2.15)**. This system is most helpful to fire suppression personnel who are operating in a building, particularly in high-rise structures that interfere with portable radio transmis-

Figure 2.15 Emergency phones allow emergency responders to communicate with the fire command center.

sions. Emergency phones are connected in the stairwells and other locations as required by the AHJ, NFPA® 72, or the Building Code. These phones enable firefighters to communicate with the Incident Commander at the Fire Command Center. Most building codes require these systems in high-rise structures.

Other Emergency Communication Systems

The International Fire Code requires new and existing buildings to be provided with approved radio coverage for emergency responders. Radio coverage inside of these buildings must be equivalent to the existing public safety communication capabilities outside of the building. Fire department radio transmissions may be augmented by the installation or use of systems. These systems may involve portable or fixed radio repeaters, bi-directional amplifiers, or a leaky coax. Radio repeaters operate by boosting or relaying fire department radio signals in buildings that may shield or disrupt normal high-frequency radio transmissions due to the weakness of these higher frequencies. Leaky coax systems are similar. However, rather than boosting or relaying radio signals, these systems simply increase the transmitting capability in these building types by creating a more effective (virtual) antenna that can improve radio communications.

Mass Notification Systems. The purpose of a **mass notification system (MNS)** is to provide emergency communications to a large number of people on a wide-scale basis **(Figure 2.16, p. 42)**. This communication can be to the occupants of a building or even an entire community. The events of September 11, 2001, as well as school shootings and other incidents, have provided evidence of the need for this type of system.

While the military was the first to implement this technology, today public and private facilities use it. Mass notification systems may be incorporated into an emergency communications system. Those individuals designing this

Mass Notification System (MNS) — System that notifies occupants of a dangerous situation and allows for information and instructions to be provided.

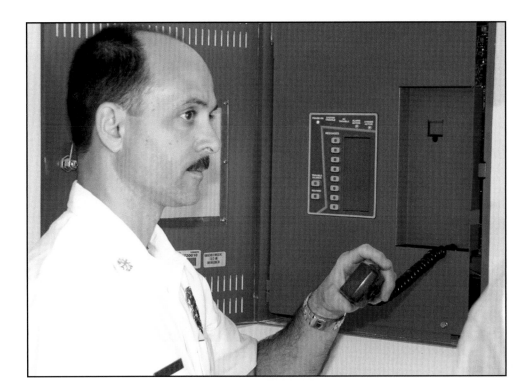

Figure 2.16 Personnel can use mass notification systems to provide emergency communications to a large number of people.

type of system must take into consideration the building being protected, as well as the needs of the occupants. When installed, mass notification systems may have a higher priority and override the fire alarm based on risk analysis. Specifications for mass notification systems are included in NFPA® 72 and should be consulted for more information.

Methods of Signal Transmission

Three methods of transmitting fire alarm signals are acceptable: wire, wireless, and fiber optic. Wire signal transmission is the oldest and utilizes copper (wire) conductors to carry the signal through the circuit between components and the fire alarm control unit.

Wired

Wire signal transmission can either be a basic one-way signal typical with conventional alarm devices or complex utilizing two-way analog, addressable, or multiplex signals as a method of transmitting signals between the FACU and components. Most common in smaller, less complex fire alarm systems, the conventional method of transmitting signals to the fire alarm control unit is similar to the operation of a light switch where "on" (closed circuit) activates the alarm system and "off" (open circuit) maintains its normal (standby) condition. The number of devices per fire alarm zone shall not exceed the maximum square footage or signaling capacity. Larger or complex fire alarm systems utilize either analog, addressable or multiplex style signals to communicate between components and the FACU. This method allows smoke detectors to automatically adjust sensitivity to compensate for dirty conditions and provide alarm verification features to reduce unwarranted alarms. In addition, the programming of each component can identify the specific location of each device, allowing circuits to maximize their capacity without being restricted to a specific geographical zone.

Wireless

Wireless utilizes radio frequencies to send and receive signals. These signals are sent between local programmable initiating devices and the protected premise FACU or a remote receiving station and the protected premise or proprietary FACU.

Fiber Optic

Fiber optics is initially the most expensive method of transmitting signals, but has long term advantages in having a larger band width, low signal loss rate, smaller size, and lighter weight when compared to copper wire. Fiber optics is typically the method of choice when networking between multiple protected premise FACUs or proprietary FACUs to the main control unit or proprietary supervising station.

Methods of Emergency Response Notification and Alarm Identification within the Protected Premise

Upon arrival at an activated fire alarm system location, the first-arriving fire official should identify the type and location of the emergency by either audible or visual means. Earlier fire alarm systems produced an audible code to identify the type and location of the activated fire alarm device, whereas newer fire alarm systems utilize annunciators or alphanumeric liquid crystal displays (LCDs) **(Figure 2.17)**. All new fire alarm systems having more than one floor or a floor area greater than 22,500 ft² (2 090 m²) require a zoned indicator panel.

Figure 2.17 An annunciator panel indicates the location of an activated alarm.

Coded

The coded method transmits the location of an activated initiating device by a series of uniform audible strokes and/or visual flashes, repeated a minimum of three times. Coded devices contain a specific numbering sequence that allows a series of strokes, flashes, and pauses to identify the floor, zone, and type of device. As an example, a series of three strokes followed by a pause, followed by one stroke, followed by a pause, followed by one stroke — 3-1-1 may indicate 3rd floor — Zone 1 — manual pull station.

Annunciators

Annunciators can either be a lamp-type that illuminates the device type, floor, and zone lamps (LEDs), or they can also be a graphic type that displays the typical floor plan with backlit or surface-mounted lamps (LEDs) that illuminates the device type, floor and zone.

Alphanumeric LCD

This device is available with programmable fire alarm systems. The activated device is displayed in text to indicate the device type, floor, and location. The major advantage to alphanumeric LCD is its ability to identify the specific location of the alarm without having to search an entire floor or zone.

Manual Fire Alarm Box

Manual alarm-initiating devices, commonly called **manual pull stations** or pull boxes, are placed in structures to allow occupants to manually initiate the fire signaling system. Manual pull stations may be connected to systems that sound local alarms, off-premise alarm signals, or both.

Although manual pull stations come in a variety of shapes and sizes, fire alarm pull stations are required to be red in color with white lettering that specifies what they are and how they are to be used. The manual pull station should only be used for fire signaling purposes unless it is designed for other uses, such as communicating with a guard station or activating a fixed fire suppression system.

According to NFPA® 72, pull stations should be mounted on walls or columns so that the operable part is no less than 3½ feet (1 m) and no more than 4 feet (1.2 m) above the floor so that all occupants can easily access it. The manual pull station should be positioned so that it is in plain sight and unobstructed. Multistory facilities should have at least one pull station on each floor. In all cases, travel distances to the manual pull station should not exceed 200 feet (60 m).

NFPA® 72 also requires that pull stations be placed within 5 feet (1.5 m) of every exit so that facility occupants can activate an alarm while they are exiting the facility **(Figure 2.18)**. Most building codes do not require manual alarm-initiating devices in structures that have automatic sprinkler systems and a device that sounds a local alarm when water begins to flow. In this situation, only one pull station is required at an approved location.

Manual pull stations that require the operator to break a small piece of glass with a mallet are no longer recommended. These devices were designed to discourage false alarms and were somewhat effective for that purpose. However, broken glass presents an injury hazard to the operator at a time when an untrained operator is least capable of clear thinking. Polycarbonate covers have taken the place of glass; however, these older types of pull stations may still be found in many old structures.

An approved wire basket or plastic cover may protect a manual pull station in areas where it would be subject to damage or accidental activation **(Figure 2.19)**. This protective device may be found in gymnasiums, materials handling areas, or in other locations where accidental activation is possible. Some pull stations leave a dye or ultraviolet residue on the activator that will transfer to the fingers of someone pulling the alarm to discourage malicious false alarms.

Manual pull stations can be single-action or dual-action. Single-action stations operate upon a single motion made by the user. When the station lever is pulled, a lever or other movable part is moved into the alarm position and a corresponding signal is sent to the FACU **(Figure 2.20)**. A dual-action station requires the operator to perform two steps in order to initiate the alarm. First,

Figure 2.18 NFPA® 72 requires that pull stations are placed within 5 feet (1.5 m) of every exit so that occupants can activate an alarm when they exit the building.

Figure 2.19 Pull stations are often equipped with protective covers to prevent damage or accidental activation. *Courtesy of Potter Signal.*

the operator must lift a cover or open a door to access the alarm control. Once this action is taken, the alarm level, switch, or button must be operated to send the signal to the FACU. Dual-action manual pull stations may be confusing to certain occupants or operators due to the need to perform two separate steps before an alarm is initiated.

Automatic Alarm-Initiating Devices

Automatic alarm-initiating devices, commonly called *detectors*, continuously monitor the atmosphere of a building, compartment, or area. When certain changes in the atmosphere are detected, such as a rapid rise in heat, the presence of smoke, or a flame signature, a signal is sent to the FACU. These devices are typically accurate at sensing the presence of the products of combustion that they are designed to detect.

Figure 2.20 A pull station in the alarm position.

Figure 2.21 A welder's arc can cause accidental activation of the fire detection system.

Fire detection and alarm system designers must take into account the normal activities and environments that occur in any given protected area. Products of combustion may be present when there is no emergency condition. Examples would be a welder's arc in an area protected by a flame detector or smoke detectors placed in an area where there is dust or excessive moisture **(Figure 2.21)**. These conditions can cause accidental activations of the system. Four basic types of automatic alarm-initiating devices are those that detect heat, smoke, fire gases, and flames. Devices that are a combination of these basic types are also available. The following sections describe these devices.

Fixed-Temperature Heat Detectors

Fire detection and alarm systems using heat-detection devices are among the oldest still in service. **Fixed-temperature heat detectors** are relatively inexpensive compared to other types of systems and are the least prone to false activations. While these devices are reliable, they are typically the slowest to activate under fire conditions. These devices also are not typically resettable and must be replaced after activation.

To be effective, heat detectors must be placed in accordance with their listing. Heat detectors must also be selected at a temperature rating that will give at least a small margin of safety above the normal ceiling temperatures that can be expected in a particular area. According to NFPA® 72, heat-sensing fire detectors must be color-coded and marked with their listed operating temperatures. For more information on detector markings, refer to NFPA® 72. All heat-detection devices operate on one or more of the three following principles:

- Heat causes expansion of various materials.
- Heat causes melting of certain materials.
- Heated materials have thermoelectric properties that are detectable.

Fire detection and alarm systems use a variety of fixed-temperature devices or detectors. The main types are the fusible link/bimetallic heat detector and continuous-line heat detector.

Fusible Elements

Fusible elements are used to hold a spring device in the detector in the open position **(Figure 2.22)**. When the melting point of the fusible link is reached, it melts and drops away, causing the spring to release and touch an electrical contact that completes a circuit and sends an alarm signal. In order to restore the detector, the fusible link must be replaced.

Bimetallic

A **bimetallic** heat detector uses two types of metal that have different heat expansion ratios **(Figure 2.23)**. Each metal is formed into thin strips, and the different metals are then bonded together. One metal expands faster than

Fixed-Temperature Heat Detector — Temperature-sensitive device that senses temperature changes and sounds an alarm at a specific point, usually 135°F (57°C) or higher.

Fusible Element — Dissimilar metals that fuse or melt when exposed to heat and allows the circuits to open or close and transmit a signal to a FACU; commonly found in fixed temperature heat detectors.

Bimetallic — Strip or disk composed of two different metals that are bonded together; used in heat-detection equipment.

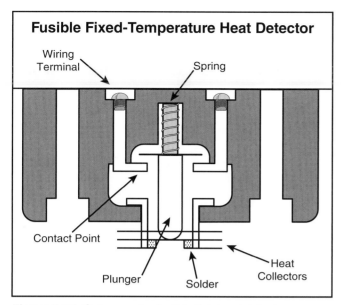

Fusible Fixed-Temperature Heat Detector

Wiring Terminal

Spring

Contact Point

Plunger

Solder

Heat Collectors

Figure 2.22 Cutaway of a fusible link heat detector.

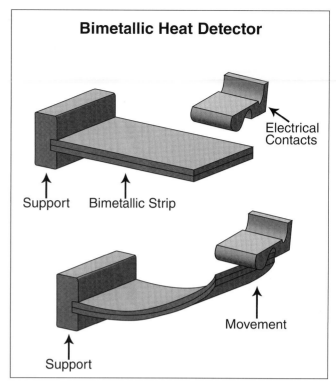

Bimetallic Heat Detector

Electrical Contacts

Support   Bimetallic Strip

Movement

Support

Figure 2.23 A bimetallic heat detector in the normal and activated positions.

the other and causes the combined strip to arch when subjected to heat. The amount that it arches depends on the characteristics of the metals, the amount of heat to which they are exposed, and the degree of arch present when in normal position. All of these factors are calculated into the design of the detector.

A bimetallic strip may be positioned with either one or both ends secured in the device. When positioned with both ends secured, a slight bow is placed in the strip. When heated, the expansion causes the bow to snap in the opposite direction. Depending on the device's design, this action either opens or closes a set of electrical contacts that, in turn, sends a signal to the FACU. Most bimetallic detectors are the automatic resetting type. They need to be checked, however, to ensure that they have not been damaged.

Continuous-Line

Most of the detectors described in this chapter are the spot style; that is, they detect conditions only at the spot where they are located. However, one style of heat-detection device, the continuous-line (or linear) device, can be used to detect conditions over a wide area. Continuous-line (linear) heat detection operates by either an increase in electrical resistance through a circuit or a short in the circuit.

Two models of continuous-line heat electrical detectors are available today. One model consists of a conductive metal inner core cable that is sheathed in stainless steel tubing. The inner core and sheath are separated by an electrically insulating semiconductor material, which keeps the core and sheath from touching but allows a small amount of current to flow between them **(Figure 2.24)**. The insulation is designed to lose some of its electrical resistance capabilities at a predetermined temperature anywhere along the line. When the heat at any given point reaches the resistance-reduction point of the

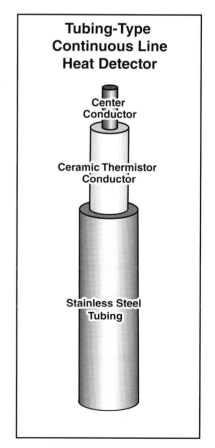

Tubing-Type Continuous Line Heat Detector

Center Conductor

Ceramic Thermistor Conductor

Stainless Steel Tubing

Figure 2.24 One example of a continuous-line heat detector.

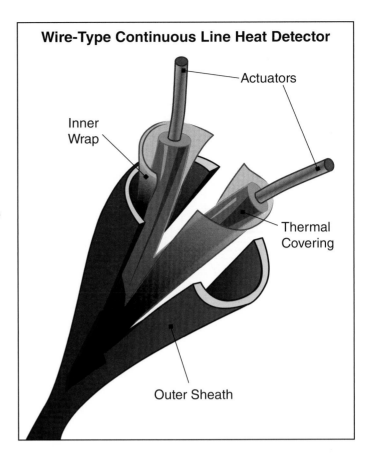

Figure 2.25 The second type of a continuous-line heat detector uses two wires that are each insulated and bundled within an outer covering.

Wire-Type Continuous Line Heat Detector

Actuators

Inner Wrap

Thermal Covering

Outer Sheath

insulation, the amount of current transferred between the two components increases. This increase results in an alarm signal being sent to the FACU. This heat-detection device restores itself when the level of heat is reduced.

A second model of continuous-line heat electrical detection devices uses two wires that are each insulated and bundled within an outer covering **(Figure 2.25)**. When the melting temperature of each wire's insulation is reached, the insulation melts and allows the two wires to touch, completing the circuit sending an alarm signal to the FACU. To restore this continuous-line heat detector, the fused portion of the wires must be removed and replaced with new wires.

Rate-of-Rise Heat Detector

A **rate-of-rise heat detector** operates on the principle that fires rapidly increase the temperature in a given area. These detectors respond in substantially lower temperatures than fixed-temperature detectors. Typically, rate-of-rise heat detectors are designed to send a signal when the rise in temperature exceeds 12°F to 15°F (7°C to 8°C) per minute because temperature changes of this magnitude are unexpected under normal, nonfire circumstances.

With proper installation, most rate-of-rise heat detectors are reliable and false activation is rare. Avoid improper placement of a heat detector to prevent false activations.

Rate-of-rise heat detectors are designed to automatically reset. Varieties of rate-of-rise heat detectors include the following:

- **Pneumatic rate-of-rise line heat detector** — Monitors large areas of a building **(Figure 2.26)**. Pneumatic line-type heat detection consists of

Rate-of-Rise Heat Detector — Temperature-sensitive device that sounds an alarm when the temperature changes at a preset value, such as 12°F to 15°F (7° to 8°C) per minute.

metal tubing arranged over a wide area of coverage. The space inside the tubing acts as a pressurized air chamber that allows the contained air to expand as it heats. These heat detectors contain a flexible diaphragm that responds to the increase in pressure from the tubing. When an area being served by the tubing experiences a temperature increase, the air pressure increases and the heat-detection device operates. The tubing in these systems is limited to about 1,000 ft (300 m) in length. The tubing should be arranged in rows that are not more than 30 ft (9 m) apart and 15 ft (4.5 m) from walls.

- **Pneumatic rate-of-rise spot heat detector** — Operates on the same principle as the pneumatic rate-of-rise line heat detector. The major difference between the two is that the spot heat detector is self-contained in one unit that monitors a specific location **(Figure 2.27)**. Alarm wiring extends from the detector back to the FACU.

- **Rate-compensation heat detector** — Contains an outer bimetallic sleeve with a moderate expansion rate. These heat detectors are designed for use in areas that are subject to regular temperature changes but at rates that are slower than those of fire conditions **(Figure 2.28)**. This outer sleeve contains two bowed struts that have a slower expansion rate than the sleeve. The bowed struts have electrical contacts. In the normal position, these contacts do not touch. When the detector is heated rapidly, the outer sleeve

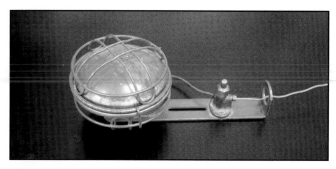

Figure 2.26 A pneumatic rate-of-rise line heat detector.

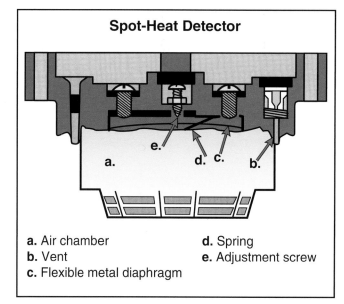

a. Air chamber   **d.** Spring
b. Vent   **e.** Adjustment screw
c. Flexible metal diaphragm

Figure 2.27 Components of a spot-heat detector.

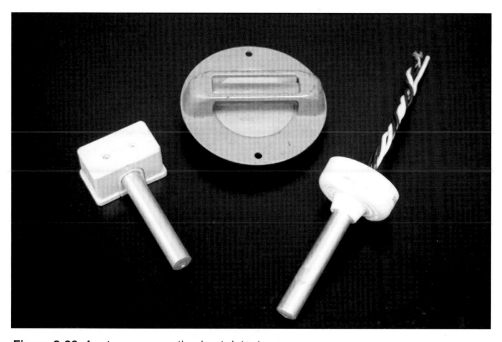

Figure 2.28 A rate-compensation heat detector.

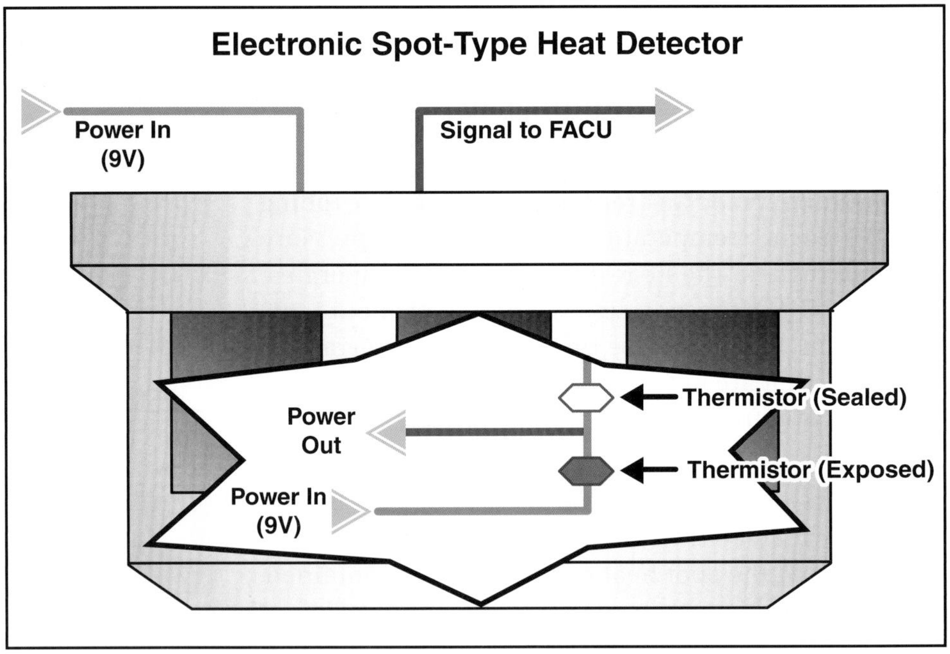

Electronic Spot-Type Heat Detector

Power In (9V)

Signal to FACU

Power Out

Power In (9V)

Thermistor (Sealed)

Thermistor (Exposed)

Figure 2.29 The rate at which the temperature increases in the internal thermistors determines the amount of current that is generated to activate and send an alarm signal.

expands lengthwise. This expansion reduces the tension on the inner strips and allows the contacts to meet, sending an alarm signal to the FACU. If the rate of temperature rise is fairly slow, such as 5 to 6°F (2°C to 3°C) per minute, the sleeve expands at a slow rate that maintains tension on the inner strips. This tension prevents unnecessary system activations.

- **Electronic spot-type heat detector** — Consists of one or more thermistors that produce a marked change in electrical resistance when exposed to heat **(Figure 2.29)**. The rate at which thermistors are heated determines the amount of current that is generated. Greater changes in temperature result in larger amounts of current flowing and activation of the alarm system. These heat detectors can be calibrated to operate as rate-of-rise detectors and function at a fixed temperature. Heat detectors of this type are designed to bleed or dissipate small amounts of current, reducing the chance of a small temperature change activating an alarm.

Smoke Detectors

Most people are well aware of the dangers of fire but less aware of the dangers of smoke inhalation. About 65 percent of fire deaths are attributed to smoke inhalation and not to burns. Smoke and toxic gases spread farther and faster than the heat from flames. When people are asleep, toxic fumes can quickly send victims into a deeper level of unconsciousness. Because of these dangers, an early warning can mean the difference between a safe escape and no escape at all.

From a life-safety standpoint, **smoke detectors** are the preferred devices in occupancies such as residences, health care facilities, and institutional care facilities. The reason is because smoke detectors sense the presence of products of combustion much more quickly than heat-detection devices **(Figure 2.30)**. Many factors affect the performance of smoke detectors, such as the type and amount of combustibles, the rate of fire growth, the detector's proximity to the fire, and ventilation within the area involved.

Smoke Detector — Alarm-initiating device designed to actuate when visible or invisible products of combustion (other than fire gases) are present in the room or space where the unit is installed.

Figure 2.30 A residential smoke detector.

Smoke detectors are tested and listed based on their performance by a third-party testing service. Regardless of their principle of operation, all smoke detectors are required to respond to the same fire tests. Two basic styles of smoke detectors are in use: photoelectric and ionization. The following sections also describe other smoke detectors such as duct, air-sampling, and video-based.

Photoelectric

A **photoelectric smoke detector** works satisfactorily on all types of fires and usually responds more quickly to smoldering fires than ionization smoke detectors. Photoelectric smoke detectors are best suited for living rooms and bedrooms because these rooms often contain large pieces of furniture, such as sofas, chairs, and mattresses, that can burn slowly and create more smoke than flames. Photoelectric smoke detectors are self-restoring after smoke dissipates and the fire alarm system is reset. A photoelectric smoke detector consists of a photoelectric cell coupled with a specific light source. The photoelectric cell functions in one of two ways to detect smoke: projected beam application (obscuration) or refractory application (scattered).

The projected-beam, or light obscuration style of photoelectric detector, uses a beam of light focused across the area being monitored onto a photoelectric-receiving device, such as a photodiode **(Figure 2.31)**. The cell constantly converts the beam into current, which keeps a switch open. When smoke interferes with or obscures the light beam, the amount of current produced is lessened. The detector's circuitry senses the change in current and initiates an alarm when a current change threshold is crossed.

Projected-beam application smoke detectors are particularly useful in buildings where a large area of coverage is desired, such as in churches, atriums, or warehouses. The projected-beam application smoke

Photoelectric Smoke Detector — Type of smoke detector that uses a small light source, either an incandescent bulb or a light-emitting diode (LED), to detect smoke by shining light through the detector's chamber: smoke particles reflect the light into a light-sensitive device called a photocell.

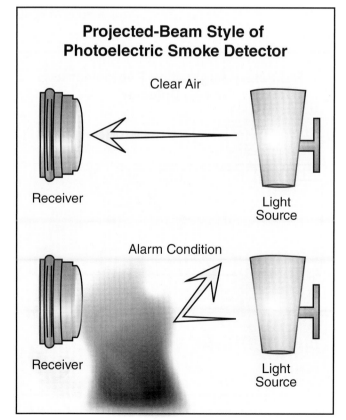

Projected-Beam Style of Photoelectric Smoke Detector

Clear Air

Receiver

Light Source

Alarm Condition

Receiver

Light Source

Figure 2.31 Principle of a projected beam-style photoelectric smoke detector.

detector is located to detect the products of combustion and transmit a signal to the FACU. Projected-beam application smoke detectors should be mounted on a stable stationary surface. Any movement due to temperature variations, structural movement, and vibrations can cause the light beams to misalign.

A refractory application, or light-scattering smoke detector, uses a beam of light from a light-emitting diode (LED) that passes through a small chamber at a point distant from the light source. Normally, the light does not strike the photocell or photodiode. When smoke particles enter the light beam, light strikes the particles and reflects in random directions onto the photosensitive device, causing the detector to generate an alarm signal **(Figure 2.32)**.

Ionization

An **ionization smoke detector** contains a sensing chamber that consists of two electrically charged plates (one positively charged and one negatively charged) and a radioactive source for ionizing the air between the plates **(Figure 2.33)**. A small amount of radioactive Americium 241 that is adjacent to the opening of the chamber ionizes the air particles as they enter. The ionized particles free electrons from the negative electron plate, and the electrons travel to the positive plate. As a result, a small ionization current measurable by electronic circuitry flows between the two plates.

Products of combustion, which are much larger than the ionized air molecules, enter the chamber and collide with the ionized air molecules. As the two interact, they combine and the total number of ionized particles is reduced. This action results in a decrease in the chamber current between the plates. An alarm is initiated when a predetermined threshold current is crossed.

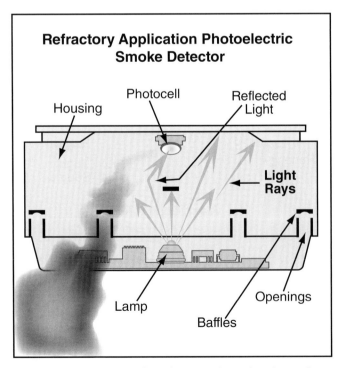

Figure 2.32 Principle of a refractory photoelectric smoke detector.

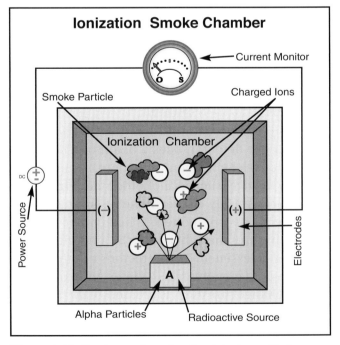

Figure 2.33 Products of combustion interfere with the ionized particles and result in a decrease in the chamber current between the plates, causing an alarm to sound.

Changes in humidity and atmospheric pressure in the room can cause an ionization detector to malfunction and initiate a false alarm. To compensate for the possible effects of humidity and pressure changes, a dual-chamber ionization detector that uses two ionization chambers has been developed and may be found in many jurisdictions. One chamber senses particulate matter, humidity, and atmospheric pressure. The other chamber is a reference chamber that is partially closed to outside air and affected only by humidity and atmospheric pressure. Both chambers are monitored electronically, and their outputs are compared. When the humidity or atmospheric pressure changes, both chambers respond equally to the change but remain balanced. When particles of combustion enter the sensing chamber, its current decreases while the reference chamber remains unchanged. The imbalance in current is detected electronically and an alarm is initiated.

An ionization smoke detector works satisfactorily on all types of fires, although it generally responds more quickly to flaming fires than photoelectric smoke detectors. The ionization detector is an automatic resetting type and is best suited for rooms that contain highly combustible materials, such as cooking fat/grease, flammable liquids, newspapers, paint, and cleaning solutions.

Duct

Duct smoke detectors are installed in the return or supply ducts or plenums of HVAC systems to prevent smoke and products of combustion from being spread throughout the building. Duct smoke detectors are specifically listed for installation within higher air velocities. Upon the detection of smoke, the HVAC system will either shut down or transition into a smoke-control mode. The detection of smoke in duct areas is sometimes difficult because the smoke can be diluted by the return air from other spaces or outside air. Duct smoke detectors are no substitute for other types of smoke detectors in open areas.

Air-Sampling

An air-sampling smoke detector is a specialty type of smoke detector that is designed to continuously monitor a small amount of air from the protected area **(Figure 2.34, p. 54)**. The cloud chamber is a type of air-sampling detector. This detector uses a small air pump to draw sample air into a high-humidity chamber within the detector. The detector then imparts the high humidity to the sample and lowers the pressure in the test chamber **(Figure 2.35, p. 54)**. Moisture condenses on any smoke particles in the test chamber, which creates a cloud inside the chamber. These detectors can be programmed for several threshold levels as well as wide sensitivity ranges. The detector triggers an alarm signal when the density of this cloud exceeds a predetermined level.

The continuous air-sampling smoke detector is composed of a system of pipes spread over the ceiling of the protected area. A fan in the detector/controller unit draws air from the building through the pipes. A photoelectric sensor then samples the air.

The spot-type aspirating smoke detector combines the spot-type photoelectric smoke detector with filtered, periodic air sampling. These detectors are designed for use in dusty areas where regular spot-type detectors cannot be used.

Figure 2.34 An air-sampling smoke detector control panel. *Courtesy of the Oak Ridge National Laboratory, U.S. Department of Energy.*

Figure 2.35 The cloud-chamber type detector is the most common air-sampling detector.

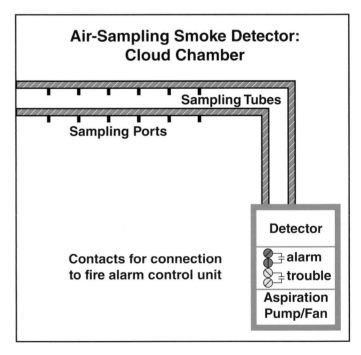

Video-Based

Video-based smoke detectors operate on the principle of detecting changes in a digital video image from a camera or a series of cameras. A closed-circuit television transmits images to a computer that looks for changes in the images. These cameras will work only in a lighted space. They also provide an image to an operator who may be monitoring the system. These systems offer advantages in large, open facilities where there may be a delay in smoke movement and detection.

Flame Detectors

A **flame detector** is sometimes called a *light detector*. There are three basic types of flame detectors:

- Those that detect light in the ultraviolet wave spectrum (UV detectors) **(Figure 2.36)**

- Those that detect light in the infrared wave spectrum (IR detectors)

- Those that detect light in UV and IR waves **(Figure 2.37)**

While these types of detectors are among the fastest to respond to fires, they may be tripped by such nonfire conditions as welding, sunlight, and other bright light sources. Flame detectors must only be placed in areas where these false triggers can be avoided or limited. They must also be positioned so that they have an unobstructed view of the protected area. If flame detectors are blocked, they cannot activate.

To prevent accidental activation from infrared light sources other than fires, an infrared detector requires the flickering action of a flame before it activates to send an alarm. This detector is typically designed to respond to a specific-sized fire from a distance determined by the manufacturer.

There are also video-based flame detectors that work on the same principle as the video-based smoke detectors. The images

Flame Detector — Detection device used in some fire detection systems (generally in high-hazard areas) that detect light/flames in the ultraviolet wave spectrum (UV detectors) or detect light in the infrared wave spectrum (IR detectors).

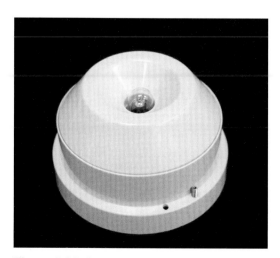

Figure 2.36 A typical UV flame detector.

Figure 2.37 A combination UV and IR flame detector.

from the closed-circuit televisions are sent to a computer with software designed to detect the characteristics of a flame. This type of flame detection system may be seen in certain chemical or petroleum facilities.

Fire-Gas Detectors

When a fire ignites in a confined area, it drastically changes the chemical-gas content of the atmosphere in the area. Some of the by-products released by a fire may include the following:

- Water vapor
- Carbon dioxide
- Carbon monoxide
- Hydrogen chloride
- Hydrogen cyanide
- Hydrogen fluoride
- Hydrogen sulfide

Only water, carbon monoxide, and carbon dioxide are released from all carbonaceous materials that burn. Whether other gases are released depends on the specific chemical makeup of the fuel, so it is only practical to monitor levels of carbon monoxide and carbon dioxide for fire detection purposes.

Fire-gas detectors implement semiconductors or catalytic elements, which are not used as frequently in other types of detectors. Fire-gas detectors can be found in such places as refineries, chemical plants, and areas of electronic assembly **(Figure 2.38)**.

Fire-Gas Detector — Device used to detect gases produced by a fire within a confined space.

Combination Detectors

Depending on the system's design, various combinations of the previously described detection devices can be incorporated in a single device.

These combinations include:

- Fixed-rate/rate-of-rise detectors
- Heat/smoke detectors
- Smoke/fire-gas detectors
- Smoke/carbon monoxide detectors

Combination detectors allow the benefit of both services and increase their responsiveness to fire conditions.

Waterflow Devices

The **waterflow device** is an automatic initiating device usually provided on automatic sprinkler systems. This device is designed to activate when water begins to flow through the sprinkler system. Modern designs include electronic flow switches that transmit signals to the FACU which in turn causes the signaling devices to function.

Waterflow Device — Initiating device that recognizes movement of water within the sprinkler or standpipe system. Once movement is noted, the waterflow device activates a local alarm and/or may transmit a signal to the FACU.

Supervisory Devices

Supervisory devices are used to supervise automatic sprinkler systems and monitor the condition of the systems. These devices monitor fire protection control valves that supply the sprinkler and other fire protection systems

(Figure 2.39). If the main water shutoff valve is closed, a supervisory indication is displayed on the FACU. Supervisory devices are also used to monitor air pressure in dry-pipe sprinkler systems, room and water temperature, tank levels, and other devices that may affect the operation of the fire protection system.

Figure 2.38 A fire gas detector.

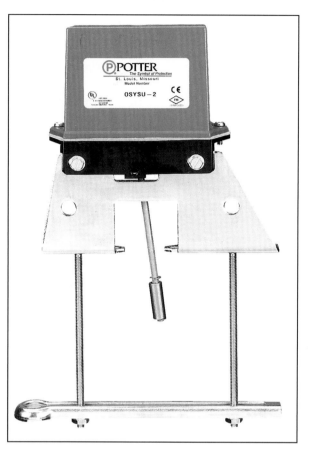

Figure 2.39 Supervisory devices are designed to monitor control valves for the protection system. *Courtesy of Potter Signal.*

Inspection and Testing of Fire Detection and Alarm Systems

To ensure operational readiness and proper performance, fire detection and alarm systems must be tested when they are installed and again on a continuing basis. Tests that are conducted when systems are installed are commonly called **acceptance tests**. Periodic testing is often referred to as **service testing**.

Fire department and fire brigade personnel who routinely conduct inspections need to have a working knowledge of these systems; however, these personnel are generally limited to visual inspections and supervision of system tests. They will typically not have to operate or maintain these systems. In most cases, company representatives or alarm system contractors actually perform system tests and maintenance. Personnel performing inspection, testing, and maintenance should be qualified and experienced in the types of devices and systems with which they work. Individuals needing more information should consult NFPA® 72.

Acceptance Test — Preservice test on fire protection, detection, and/ or suppression systems after installation to ensure that the system operates as intended.

Service Test — Series of tests performed on fire protection, detection, and/ or suppression systems in order to ensure operational readiness. These tests should be performed at least yearly or whenever the system has undergone extensive repair or modification.

Acceptance Testing

Acceptance testing is performed soon after the system has been installed or modified to ensure that it meets design criteria and functions properly. The occupant's insurance carrier and/or local codes and ordinances may require acceptance tests. Representatives of the building owner or occupant, the fire department, and the system installer/manufacturer should witness acceptance tests. The fire department representative may be a fire inspector, a fire protection engineer, or in some cases the fire marshal.

Some jurisdictions require the system installer or manufacturer to document that the system is ready for inspection by the AHJ. This record prevents the fire inspector or engineer from checking a system that is not ready for acceptance testing.

It is important to test all components of the alarm system. All of the functions of the fire detection and alarm system should be operated and the following tasks performed during acceptance tests:

- **Alarm and trouble modes of system operation** — Check actual wiring and circuitry against the system drawing to ensure that all are connected properly.

- **Fire alarm control unit (FACU)** — Operate all interactive controls at the FACU to ensure that they control the system as designed. Inspect thoroughly to ensure that the FACU is in proper working order **(Figure 2.40)**.

Figure 2.40 The FACU should be inspected to ensure it is in proper working order.

- **Alarm-initiating and signaling devices and circuits** — Check all items for proper operation. Test pull stations, detectors, bells, strobe lights, and other devices to ensure that they are operational. Test each initiating device to ensure that it sends an appropriate signal and causes the system to go into the alarm, supervisory, or trouble mode.

- **Power supplies** — Operate the system on both the primary and secondary power supplies to ensure that both will supply the system adequately.

Restorable heat detectors should be inspected by adhering to approved testing procedures described by the manufacturer. Nonrestorable heat detectors must not be tested during field inspections.

Some combination detectors have both restorable and nonrestorable elements. Caution must be exercised to avoid tripping the non-restorable element. Nonrestorable pneumatic detectors should be tested mechanically. Those detectors equipped with replaceable fusible elements should have the elements removed to see whether the contacts touch and send an alarm signal. The elements can be replaced following the test.

The manufacturers of smoke, flame, and fire-gas detectors usually have specific instructions for testing their detectors. These instructions must be followed on both the acceptance and service tests. Testing may include the use of smoke-generating devices, aerosol sprays, or magnets. The use of non-approved testing devices may result in the manufacturer's warranty on the detector being voided.

To respond to an alarm, it is also important to determine the ability of outside entities, such as the central station, auxiliary alarm systems, remote stations, and proprietary systems. The alarm receiving capability must be verified, and a telecommunication facility must ensure that the signals are properly received.

The results of all tests must be documented to the satisfaction of both the insurance carrier and the fire department. Issuing the alarm system acceptance is typically a preliminary step toward the issuance of a certificate of occupancy. NFPA® 72 contains complete information on acceptance testing.

Service Testing and Periodic Inspection

To ensure that fire detection and alarm systems work when they are needed, the systems must be tested and inspected on a regular basis. Locally adopted fire codes sometimes mandate that these tests be witnessed by members of the AHJ's inspection division. The actual performance of the tests is the responsibility of the owner/occupant or the fire alarm monitoring company. Periodic tests and inspections are performed on all components of the fire detection and alarm system, including the initiating devices and the FACU.

Fire Detection and Alarm Systems

Fire detection and alarm-signaling equipment should receive a general inspection on a routine basis. The AHJ and the building owner/occupant should conduct the inspection. Because there are many occupancies in a jurisdiction, fire department personnel may not always be available to witness every test.

Building owners/occupants are typically required to test the systems on their own and document the results. At specified intervals, fire department personnel will also be present to witness tests. The procedures addressed in this section should be applied during inspection and testing of the systems. Activities to be performed during an inspection include.

- Inspect all wiring for proper support.

- Look for wear, damage, or any other defects that may render the insulation ineffective.

- Inspect conduit for solid connections and proper support wherever circuits are enclosed.

- Check batteries that are used as an emergency power source for clean contact and proper charge.

- Ensure that all equipment, especially initiating and signaling devices, are free of dust, dirt, paint, and other foreign materials. When dust or dirt is found, devices can be cleaned with a vacuum cleaner rather than by wiping. Wiping tends to spread debris around, causing it to settle on electrical contacts. This debris may inhibit the future operation of the system.

- Ensure that access to FACUs, recording instruments, and other devices is not obstructed in any fashion and no objects stored on, in, or around these systems. Many FACUs have storage areas with locking doors for extra relays, lightbulbs, and test equipment.

Fire Alarm-Initiating Devices

Any fire detection and alarm-signaling system will be ineffective unless the alarm-initiating devices are in proper working order and send the appropriate signal to the system control panel. These devices need to be tested per applicable standards to ensure that they are operational. A representative of the AHJ, such as an inspector or engineer, can witness alarm-initiating device tests.

Numerous items need to be checked when testing and inspecting a manual alarm-initiating device. Ensure that the following tasks are performed during an inspection:

- Install each device for its proper application.

- Check that access to the device is unobstructed.

- Ensure that each unit is easy to operate **(Figure 2.41)**.

- Close the unit housing tightly to prevent dust and moisture from entering and disrupting service.

- Remove any chipped, cracked, or otherwise impaired glass and replace.

- Check that the device's cover or door opens easily and all the components behind the cover or door are in place and ready for service.

- Ensure that the components are compatible for use in the system.

Without functional fire detection devices, the most elaborate signaling systems are useless. The reliability of the entire system is, in fact, based largely on the reliability of the detection devices.

Automatic alarm-initiating devices should be checked after installation, after a fire, and at recurring times based on guidelines established by the AHJ or the manufacturer **(Figure 2.42)**. These guidelines are often found in the locally adopted fire code. All detector testing should be in accordance with local guidelines, manufacturer's specifications, and NFPA® 72. Specifically,

Figure 2.41 Manual pull stations should be tested to ensure they are easy to operate.

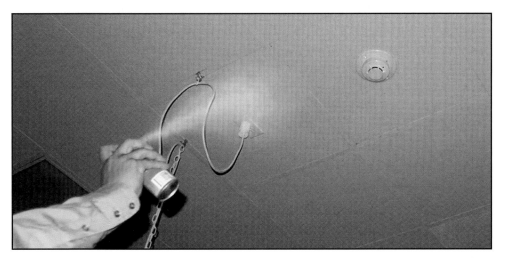

Figure 2.42 Automatic alarm-initiating devices should be tested as required by the AHJ.

different manufacturers' components may be incompatible when used in the same system.

Replace detectors found in one or more of the following conditions or send them to a recognized testing laboratory for testing:

- Systems that are being restored to service after a period of disuse
- Detectors that are obviously corroded
- Detectors that have been painted, even if attempts were made to clean them
- Mechanically damaged or abused detectors
- Circuits that were subjected to current surges, overvoltages, or lightning strikes
- Detectors subjected to foreign substances that might affect their operation
- Detectors subjected to direct flame, excessive heat, or smoke damage

The system owner should maintain a permanent record of all detector records for at least five years. Information that should be included in the record is the test date, the detector type, the detector's location, the type of test, and the test result.

A nonrestorable fixed-temperature detector cannot be tested periodically. Testing would destroy the detector and require the system to be rendered inoperable until a replacement detector could be located and installed. For this reason, tests are not required until 15 years after the detector has been installed. At that time, two percent of the detectors must be removed and laboratory tested. If a failure occurs in one of the detectors, additional detectors must be removed and laboratory-tested. These tests are designed to determine if there is a problem with failure of the product in general or a localized failure involving one or two detectors.

Periodic testing procedures are included in NFPA® 72 and in the fire alarm manufacturer's literature for the system and its components. The following periodic tests are recommended test procedures for the various devices discussed:

Figure 2.43 Fusible element detectors are tested by removing the element and observing whether or not the contacts close.

- **Restorable heat detection devices** — Test one detector on each signal circuit semiannually. Check as described previously in the section on acceptance testing. A different detector should be selected each time and so noted in the inspection report. Subsequent inspections should include a copy of the previous report to ensure that the same detector is not tested each time.

- **Fusible element detector with replaceable discs** — Check semiannually by removing the element and observing whether the contacts close **(Figure 2.43)**. After the test, the fusible element must be reinstalled. The elements should be replaced at five-year intervals.

- **Pneumatic heat detector** — Test semiannually with a heating device or a pressure pump. If a pressure pump is used, the manufacturer's instructions must be followed.

- **Smoke detector** — Test semiannually in accordance with the manufacturer's recommendations. The manufacturer usually provides the instruments required for performance and sensitivity testing. Sensitivity testing should be performed after the detector's first year of service and every two years after that.

- **Flame and gas detection devices** — Require testing by highly trained individuals because they are complicated devices. Professional alarm service technicians typically perform testing on a contract basis.

- **Fire Alarm Control Units (FACUs)** — Check to ensure that all parts are operating properly. All switches should perform their intended functions and all indicators should illuminate or sound when tested. When individual detectors are triggered, the FACU should indicate the proper location and warning lamps should light. Indicated locations could very well be out of date due to renovations.

Check auxiliary devices at this time. The auxiliary devices include local evacuation alarms and HVAC functions, such as air-handling system shutdown controls and smoke dampers. All devices must be restored to proper operation after testing. In connection with these tests, the receiving signals should also be checked. The proper signal and/or number of signals should be received and recorded. Signal impulses should be definite, clear, and evenly spaced to identify each coded signal. No sticking, binding, or other irregularities should occur.

At least one complete round of printed signals should be clearly visible and unobstructed by the receiver at the end of the test. The time stamp should clearly indicate the time of the signal and should not interfere in any way with the recording device.

System Inspection and Testing Frequencies

The following list gives a brief synopsis of the inspection and testing requirements for various types of systems and timetable guidelines. If any of these systems use backup electrical generators for emergency power, those generators should run under load monthly for at least 30 minutes.

- **Local alarm systems** — Test in accordance with guidelines established in NFPA® 72 and the manufacturer's recommendations.

- **Central station system** — Test signaling equipment on a monthly basis. Check waterflow indicators, automatic fire detection systems, and supervisory equipment bimonthly. Check manual fire alarm devices, water tank level devices, and other automatic sprinkler system supervisory devices semiannually. When these tests are scheduled, both facility/building supervisory personnel and central station personnel should be notified before the test to prevent them from evacuating occupants or dispatching fire units.

- **Auxiliary fire alarm systems** — Visually inspect and actively test monthly (by the occupant) to ensure that all parts are in working order and the operation of the system results in a signal being sent to the fire department telecommunications center. Test noncoded manual fire alarm boxes semiannually.

- **Remote station and proprietary systems** — Test according to the testing requirements established by the AHJ. Test fire detection components of these systems monthly. Test waterflow indicators semiannually; however, the frequency of testing may depend upon the type of indicator.

- **Emergency communications systems** — Conduct functional tests of the various components in these systems quarterly (by the owner/occupant). Include selected parts of the system that are reflective of what may actually be used during an incident. Check all components at least annually.

Chapter Summary

Whether a fire detection and alarm system is simple or complex, its quick operation is vital if fire emergencies are to be mitigated with as little loss or damage as possible. The earliest systems of ringing bells and crying out warnings have given way to complex electronic systems that alert occupants and emergency services at the same time. Fire fighting personnel must be familiar with the basic operating features of the fire detection and alarm systems in their response areas so that they can recognize obvious problems that may be occurring. They also need to know which types of detection systems are appropriate for certain hazards so that the systems that are installed can operate effectively. Because this level of inspection is confined to visual observations and supervision of systems testing, other trained personnel should be involved to ensure system operability and safety.

Review Questions

1. What are the basic components of fire detection and alarm systems?

2. What are the types of fire detection and alarm systems?

3. What are the three methods of transmitting fire alarm signals?

4. Identify the three methods that fire alarms use to transmit the location of an activated initiating device.

5. What is the difference between single-action and duel-action manual pull stations?

6. What are the three main types of fixed-temperature heat detectors?

7. Identify two types of rate-of-rise heat detectors.

8. How do photoelectric smoke detectors function?

9. What measures can be taken to prevent accidental and false triggers in flame detectors?

10. What are the four tasks that should be performed during acceptance tests?

11. What are three conditions in which detectors should be replaced or sent to a recognized laboratory?

12. What are the system inspection and testing requirements for central station systems?

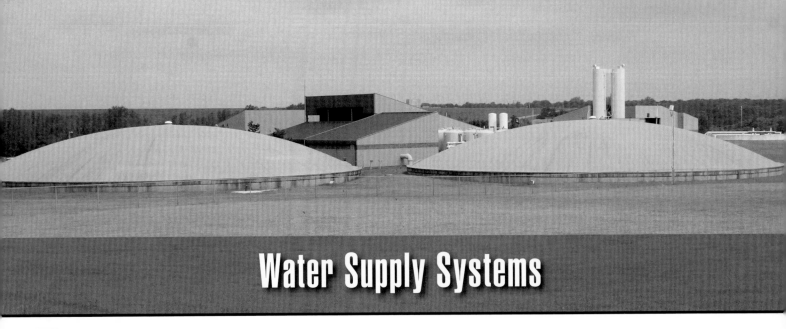

Water Supply Systems

Chapter Contents

Key Terms

FESHE Outcomes

Fire and Emergency Services Higher Education (FESHE) Outcomes: Fire Protection Systems

2. Describe the basic elements of a public water supply system including sources, distribution networks, piping, and hydrants.

3. Explain why water is a commonly used extinguishing agent.

Water Supply Systems

Learning Objectives

After reading this chapter, students will be able to:

1. Describe the properties of water as an extinguishing agent.

2. Summarize the principles of pressure.

3. Describe the different types of pressure as they relate to water.

4. Summarize the principles of friction loss in water supply systems.

5. Describe the basic components of public water supply systems.

6. Describe the basic components of private water supply systems.

7. Summarize variables that affect piping distribution systems.

8. Distinguish among valves used on water distribution systems.

9. Summarize the types of fire hydrants and their maintenance and testing procedures.

10. Identify fire hydrant locations and distribution methods.

Chapter 3
Water Supply Systems

Case History

While performing fire hydrant flow tests for a water distribution system study at a military base, engineers recorded some abnormal results in a sector of the base. After questioning personnel within the base's public works department, it was determined that there should be a valve serving that sector, even though no one knew its exact location. After extensive searching and discussion with four public works department employees, the subject valve was located. Dirt partially covered the valve. When the valve key was inserted to verify that the valve was open, the engineers discovered that the ground had shifted such that the sleeve was no longer over the valve stem and the valve could not be controlled. After crews had excavated around the valve, it was determined that the valve was in the closed position. No personnel with the public works department knew of a time when the valve had been operated within the last five years. Fortunately, there had not been a fire in that sector of the base during that time. Had a fire occurred, the fire department would have had no access to the water supply and the results could have been detrimental.

Water continues to be the primary extinguishing agent due to its availability, affordability, and effectiveness in spite of the many new innovations and techniques for controlling and fighting fires that have been developed. In addition, water is easily stored and can be conveyed long distances. For these reasons, fire service personnel must understand the process of obtaining and moving water.

Basic knowledge about water and water supply systems is necessary when individuals design, inspect, and maintain fire protection systems. The quantity of water necessary and the corresponding pressures must be determined in order for the fire protection to be adequate for a building or facility. Individuals must examine existing water supply systems to determine whether the calculated needs can be met or alternative water supplies are necessary. The water supply requirements can shift due to an occupancy change or hazard change.

It is impossible for this manual to cover all the specifics of water supply systems for each individual community. Individuals responsible for the design, installation, inspection, or maintenance of fire protection systems must determine the adequacy of local water supplies and systems. This chapter provides information on water supply systems in communities and fire and emergency services operations in the use and support of those systems.

Water as an Extinguishing Agent

Water exists in a liquid state between 32°F and 212°F (0° and 100°C). Below 32°F (0°C), water experiences a phase change to a solid state of ice. Above 212°F (100°C), it vaporizes into water vapor or steam **(Figure 3.1)**. Water cannot be seen in vapor form; it only becomes visible as it rises away from the surface of the liquid water and begins to condense.

Water is considered to be incompressible, and its weight varies at different temperatures. Water's density, or its weight per unit of volume, is measured in pounds per cubic foot (lb/ft³) (kg/m³) or pounds per gallon (lb/gal) (kg/L). Water is heaviest and has its highest density close to its freezing point. Water is lightest and has its lowest density close to its boiling point. For fire protection purposes, ordinary water is generally considered to weigh 62.4 lb/ft³ (1 000 kg/m³) or 8.34 pounds per gallon (1 kg/L). The following sections will explain these properties.

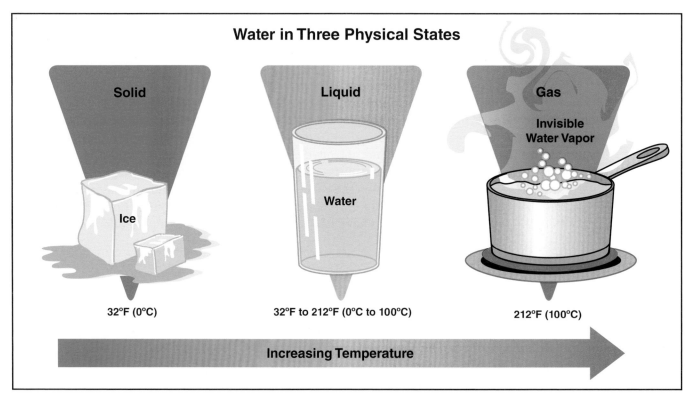

Figure 3.1 Water exists in three physical states: solid, liquid, and gas.

Extinguishing Properties of Water

Water has the ability to extinguish fire in several ways. The primary way is by cooling, namely absorbing heat from the fire. Another way is smothering, which occurs when oxygen is excluded from the fire. Smothering works especially well on the surface of burning liquids that are heavier than water. A smothering effect also occurs to some extent when water converts to steam in a confined space.

The extinguishing property of water is affected by the following four factors:

- Law of specific heat
- Law of latent heat of vaporization
- Surface area of the water
- Specific gravity

Law of Specific Heat

Water absorbs large amounts of heat. **Specific heat** is a measure of the heat-absorbing capacity of a substance, meaning the amount of heat required to change a unit quantity of a substance by one degree in temperature. The amount of heat transfer is usually measured in **British thermal units (BTU)** or in the International System of Units (SI) as joules (J). A BTU is the amount of heat required to raise the temperature of 1 pound (0.454 kg) of water 1°F (.556°C). The joule, also a unit of work, has taken the place of the calorie in the SI heat measurement.

The specific heat of different substances varies. **Table 3.1** shows various fire extinguishing agents and the specific heat comparisons of each with water. These comparisons are based upon the weight of the agent. Dividing the specific heat of water by the specific heat of another extinguishing agent demonstrates that water is clearly the best material for absorbing heat. For example, water absorbs five times as much heat as does an equal amount of carbon dioxide gas.

Table 3.1
Specific Heat of Extinguishing Agents

Agent	Specific Heat
Water	1.00
Calcium chloride solution	0.70
Carbon dioxide (solid)	0.12
Carbon dioxide (gas)	0.19
Sodium bicarbonate	0.22

Specific Heat — The amount of heat required to raise the temperature of a specified quantity of a material and the amount of heat necessary to raise the temperature of an identical amount of water by the same number of degrees.

British Thermal Unit (BTU) — Amount of heat energy required to raise the temperature of 1 lb of water 1°F. 1 BTU = 1.055 kilo joules (kJ).

Latent Heat of Vaporization — Quantity of heat absorbed by a substance at the point at which it changes from a liquid to a vapor.

Law of Latent Heat of Vaporization

The **latent heat of vaporization** is the quantity of heat absorbed by a substance when it changes from a liquid to a vapor. The temperature at which a liquid absorbs enough heat to change to vapor is known as the *boiling point*. At sea level, water begins to boil or vaporize at 212°F (100°C). Vaporization, however, does not completely occur the instant water reaches the boiling point. Each pound (0.454 kg) of water requires approximately 970 BTU (1 023 kJ) of additional energy (heat) to completely convert into steam.

NOTE: It takes 1 BTU (1.055 kJ) to raise the temperature of 1 pound (0.454 kg) of water by 1°F (0.556°C) at sea level.

The latent heat of vaporization is significant in fire fighting because the temperature of the water is not increased beyond 212°F (100°C) during the absorption of the 970 BTU (1 023 kJ) for every pound (0.454 kg) of water. For example, a gallon (3.8 L) of water weighs 8.34 pounds (3.8 kg). At 60°F (15.6 °C), it requires 152 BTU (160 kJ) to raise the temperature of each pound of water to 212°F (100°C):

Customary

$$212°F - 60°F = 152°F \times \frac{1 \text{ BTU}}{1 \text{ lb} \times 1°F} = 152 \text{ BTU/lb}$$

Metric

$$100°C - 15.6C = 84.4°C \times \frac{4.184 \text{ kJ}}{1 \text{ kg} \times 1°C} = 353 \text{ kJ/kg}$$

Therefore, 1 gallon (3.8 L) of water absorbs 1,268 BTU (1 338 kJ) of energy (heat) to get to 212°F (100°):

Customary

$$\frac{152 \text{ BTU}}{\text{lb}} \times \frac{8.34 \text{ lbs}}{1 \text{ gal}} = \frac{1268 \text{ BTU}}{\text{gal}}$$

Metric

$$353 \text{ kJ/kg} \times 1 \text{ kg} = \frac{353 \text{ kJ/L}}{1 \text{ L H}_2\text{O}}$$

Because the conversion to steam requires another 970 BTU/lb (2 253 kJ/kg), an additional 8,090 BTU (8 535 kJ) of energy (heat) will be absorbed through this process:

Customary

970 BTU/lb x 8.34 lbs/gal H_2O = 8,090 BTU/gal

Metric

2 253 kJ/kg x 1 kg/1 L H_2O = 2 253 kJ/L

This means that one gallon (3.8 L) of water will absorb 9,358 BTU (9 873 kJ) of energy (heat) if all the water is converted to steam **(Figure 3.2)**. In fire fighting terms, if water from a 100 gpm (400 L/min) fog nozzle is projected into a highly heated area, it would absorb approximately 935,800 BTU (987 325 kJ) of energy (heat) per minute if all of the water is converted to steam and the water starts at 60°F (15.6°C).

Surface Area of Water

The speed with which water absorbs heat increases in proportion to the water surface exposed to the heat. For example, if a 1-inch (25 mm) cube of ice is dropped into a glass of water, it will take a considerable amount of time for the ice cube to absorb its capacity of heat, or melt; because only 6 square inches (3 870 mm²) of ice are exposed to the water. If the same cube of ice is divided into smaller pieces, such as 1/8-inch cubes (3 mm), and they are dropped into the water, 48 inch² (30 967 mm²) of the ice are exposed. Although the smaller cubes equal the same mass of ice as the larger cube, the smaller cubes melt faster. The larger overall surface area of crushed ice causes it to melt in a drink faster than cubed ice **(Figure 3.3)**. The same theory applies to water in its liquid state. If water is divided into many drops, the rate of heat absorption increases hundreds of times.

The expansion capability of water as it converts to steam is another characteristic that aids in fire fighting. Expansion helps cool the fire by driving heat and smoke from the area. The temperature of the fire area will determine the amount of heat expansion. At 212°F (100°C), water expands approximately 1,700 times its original volume.

Specific Gravity

The density of liquids in relation to water is known as **specific gravity**. Water is given a value of 1. Liquids with a specific gravity less than 1 are lighter than water and therefore float on water. Those liquids with a specific gravity

Specific Gravity — Weight of a substance compared to the weight of an equal volume of water at a given temperature. A specific gravity less than 1 indicates a substance lighter than water; a specific gravity greater than 1 indicates a substance heavier than water.

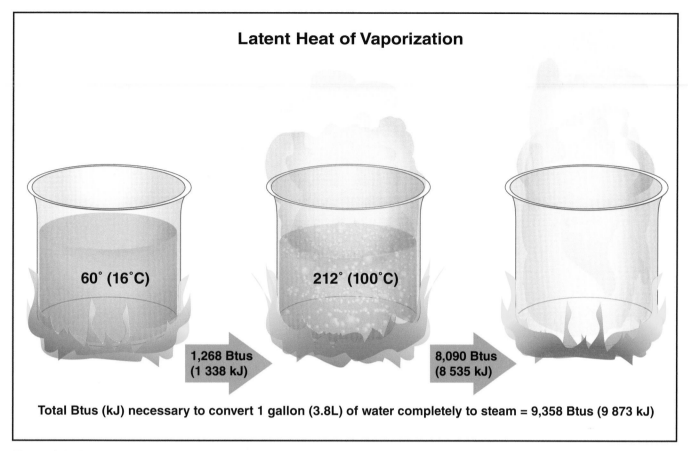

Latent Heat of Vaporization

60° (16°C)

212° (100°C)

1,268 Btus
(1 338 kJ)

8,090 Btus
(8 535 kJ)

Total Btus (kJ) necessary to convert 1 gallon (3.8L) of water completely to steam = 9,358 Btus (9 873 kJ)

Figure 3.2 An example of the latent heat of vaporization.

greater than 1 are heavier than water and will sink to the bottom. If the other liquid also has a specific gravity of 1, it mixes evenly with water. Most flammable liquids have a specific gravity of less than 1 **(Table 3.2, p. 74)**.

Water can also smother a fire when it floats on liquids that have a higher specific gravity than water, such as carbon disulfide. If the material is water soluble, such as alcohol, the smothering action is unlikely to be effective. Water may also smother fire by forming an emulsion over the surface of certain combustible liquids. When a spray of water agitates the surface of these liquids, the agitation causes the water to be temporarily suspended in emulsion bubbles on the surface. These bubbles then smother the fire. The emulsifying ability of a liquid is dependent upon other properties it possesses. Therefore, if a firefighter directs water onto a flammable liquid fire improperly, the fire can just float away on the water and ignite everything in its path. The use of fire fighting foam can control this situation because it floats on the surface of the flammable liquid and smothers the fire.

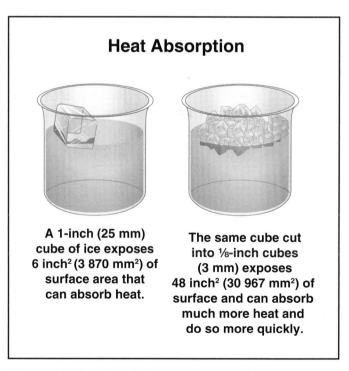

Heat Absorption

A 1-inch (25 mm) cube of ice exposes 6 inch² (3 870 mm²) of surface area that can absorb heat.

The same cube cut into ⅛-inch cubes (3 mm) exposes 48 inch² (30 967 mm²) of surface and can absorb much more heat and do so more quickly.

Figure 3.3 Even though the same amount of ice is used, more heat can be absorbed by the smaller cubes due to their increased surface area over that of the large cube.

Table 3.2
Specific Gravity of Common Substances

Substance	Specific Gravity at 68°F
Water	0.998
Commercial Solvent	0.717
Carbon Tetrachloride	1.582
Medium Lubrication Oil	0.891
Medium Fuel Oil	0.854
Heavy Fuel Oil	0.908
Regular Gasoline	0.724
Turpentine	0.862
Ethyl Fuel	0.789
Benzene	0.879
Glycerin	1.262
Light Machinery Oil	0.907
Air	0.0012
Ammonia	0.0007
Carbon Dioxide	0.0018
Methane	0.0017

Advantages and Disadvantages of Water

Some advantages of using water as an extinguishing agent include the following:

- **Heat-absorbing capacity** — Water has a greater heat-absorbing capacity than other common extinguishing agents due to the relatively large amount of heat required to change water to steam.

- **Surface area** — Water applied to a fire by fog patterns and deflected solid streams greatly increases its surface area and causes heat to be absorbed more rapidly.

- **Availability** — In many areas water is plentiful and readily available.

- **Cost** — Water is relatively inexpensive when compared to other commercially available extinguishing agents

Some disadvantages of using water as an extinguishing agent include the following:

- **High surface tension** — Water does not readily soak into dense materials. The addition of wetting agents to water can decrease this surface tension and improve penetration.

- **Reactivity** — Water is reactive with certain substances, such as combustible metals **(Figure 3.4)**.

- **Low opacity and reflectivity** — Water allows radiant heat to easily pass through it.

- **Freezing point** — Freezing water can pose a danger to suppression systems. In cold climates, ice may form in and on equipment, causing the potential for malfunction.

- **Conductivity** — Water conducts electricity, which can be a hazard when fire fighting operations are taking place around energized electrical equipment (**Figure 3.5**).

- **Contaminated runoff** — Water can carry contaminants away from the fire scene (**Figure 3.6**).

- **Weight** — Excessive water weight can contribute to structural instability.

- **Availability of Water** — May not be plentiful or available for areas with little water or areas that are drought stricken.

Figure 3.4 Placards in some storage facilities indicate that no water should be used due to its reactivity with the item being stored. *Courtesy of Rich Mahaney.*

Figure 3.5 Ice can be a hazard when fire fighting operations are taking place around energized electrical equipment. *Courtesy of Ron Jeffers.*

Figure 3.6 An example of contaminated runoff from a fire scene. *Courtesy of Ron Jeffers.*

Pressure — Force per unit area exerted by a liquid or gas measured in pounds per square inch (psi) or kilopascals (kPa).

Water Pressure

In fire service terminology, **pressure** refers to the force that moves water through a conduit — either a pipe or a hose. Pressure is defined as force per unit area in a liquid or gas. It can be expressed in pounds per square inch (psi), pounds per square foot (psf), or kilopascals (kPa).

Pressure

The United States fire service measures pressure, whether it is gaseous (air) or liquid (water) by pounds per square inch (psi). Outside of the United States, pressure is measured in either kilopascals (kPa) or bars. The International System of Units (SI) is the modern form of the metric system and is the world's most widely used system of measurement in commerce and science. SI recognizes kilopascals for the measurement of pressure. One psi equals approximately 7 kPa. Bars are another unit for pressure measurement. One bar is equal to 100 kPa.

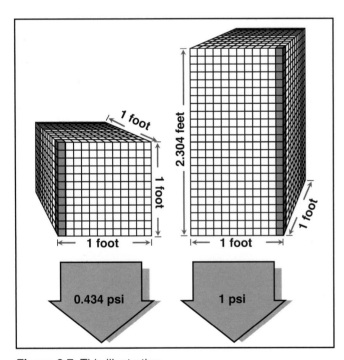

Figure 3.7 This illustration shows the relationship between height and pressure in the customary system of measurement.

In order to determine pressure, it is necessary to know the weight of water and the height that a column of water occupies. The weight of 1 cubic foot (ft³) of water is approximately 62.4 pounds. Because 1 square foot (ft²) contains 144 inches², the weight of water in a 1-inch² column of water 1-foot high equals 62.4 pounds divided by 144 inches², or 0.434 pounds. Therefore, a 1-inch² column of water 1-foot high exerts a pressure 0.434 psi at its base **(Figure 3.7)**.

In metrics, a cubic decimeter (a cube that is 0.1 m × 0.1 m × 0.1 m) holds 1 liter (L) of water. The weight of 1 liter of water is 1 kilogram (kg). The cube of water exerts 1 kilopascal (kPa) of pressure at the bottom of the cube. One cubic meter of water holds 1 000 liters of water and weighs 1 000 kilograms. Because the cubic meter of water is comprised of 100 columns of water, each column being 10 decimeters tall, each column exerts 10 kilopascals at its base **(Figure 3.8)**. Six principles determine how fluids act when pressure is applied. These principles are introduced in the following sections.

Principles of Pressure

Fire and emergency services personnel must understand the various sources of pressure. These principles are:

First Principle

Fluid pressure is perpendicular to any surface on which it acts. In a flat-sided vessel containing water, the pressure exerted by the weight of the water is perpendicular to the container walls **(Figure 3.9)**. If this pressure is exerted in any other direction, the water starts moving downward along the sides and rising in the center.

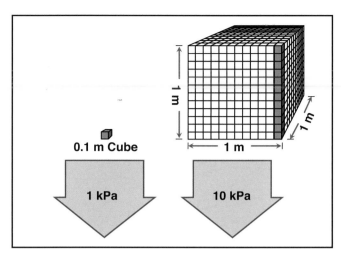

Figure 3.8 This illustration shows the relationship between height and pressure in the metric system of measurement.

Second Principle

The second pressure states that fluid pressure at a point in a fluid at rest is the same intensity in all directions. In other words, fluid pressure at a point in a fluid at rest has no direction.

Third Principle

Pressure applied to a confined fluid is transmitted equally in all directions. This principle is illustrated by viewing a hollow sphere that is attached to a water pump **(Figure 3.10)**. A series of gauges is set into the sphere around its circumference. When the sphere is filled with water and pressure is applied by the pump, all gauges will register the same pressure. This result is true if they are on the same grade line with no change in elevation.

Fourth Principle

The pressure of a liquid in an open vessel is directly proportional to its depth. This principle is illustrated by observing three vertical containers, each 1 inch² (645 mm²) with different depths of water. If the first container has 1 foot (300 mm) of water, the second 2 feet (600 mm), and the third 3 feet (900 mm), the pressure at the bottom of the second container will be twice that of the first and the pressure at the bottom of the third will be three times that of the first **(Figure 3.11, p. 78)**.

Fifth Principle

The pressure of a liquid in an open vessel is directly proportional to the density of the liquid. If one container holds mercury and the other holds water, it would take 13.55 inches (344 mm) of water to produce the same pressure as 1 inch (25 mm) of mercury. Mercury is, therefore, much denser than water **(Figure 3.12, p. 78)**.

First Principle of Pressure

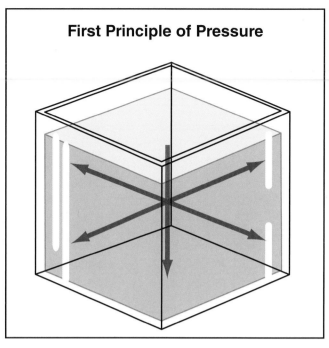

Figure 3.9 A vessel having flat sides and containing water illustrates the first principle.

Third Principle of Pressure

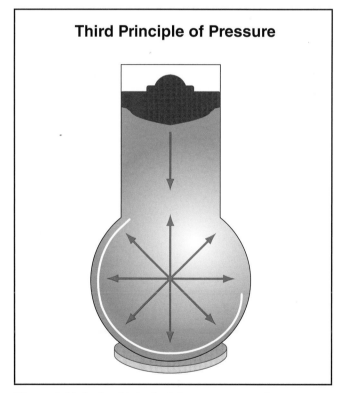

Figure 3.10 In the third principle, pressure applied to a confined fluid is transmitted equally in all directions.

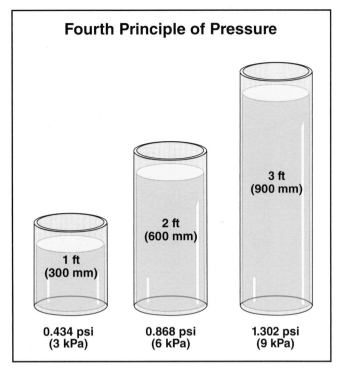

Fourth Principle of Pressure

3 ft
(900 mm)

2 ft
(600 mm)

1 ft
(300 mm)

0.434 psi
(3 kPa)

0.868 psi
(6 kPa)

1.302 psi
(9 kPa)

Figure 3.11 In the fourth principle, the pressure of a liquid in an open vessel is directly proportional to its depth.

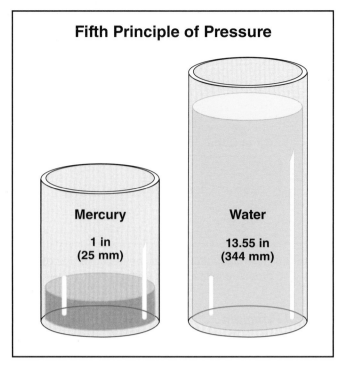

Fifth Principle of Pressure

Mercury
1 in
(25 mm)

Water
13.55 in
(344 mm)

Figure 3.12 In the fifth principle, 1 inch (25 mm) of mercury creates the same pressure at the bottom of a container as 13.55 inches (344 mm) of water.

Sixth Principle

The pressure of a liquid on the bottom of a vessel is independent of the shape of the vessel — regardless of the container's shape and its opening; the pressure of the water column is the same for a given height.

Types of Pressure

A variety of terms are used for different types of pressure that are encountered in water supply systems. Properly identifying these terms is important so that they can be understood when working with water supply systems. The types of pressure include:

● Atmospheric pressure

● Head pressure

● Static pressure

● Normal operating pressure

● Residual pressure

● Flow pressure (velocity pressure)

Atmospheric Pressure

The pressure that is exerted on the earth by the atmosphere itself is called **atmospheric pressure**. The atmosphere surrounding the earth has depth and density and exerts pressure upon everything. Atmospheric pressure is greatest at low altitudes. At sea level, atmospheric pressure is 14.7 psi (101 kPa). This is considered standard atmospheric pressure. As the altitude increases, atmospheric pressure decreases **(Figure 3.13)**. The readings on most pressure

Atmospheric Pressure — Force exerted by the atmosphere at the surface of the earth due to the weight of air. Atmospheric pressure at sea level is about 14.7 psi (101 kPa). Atmospheric pressure increases as elevation is decreased below sea level and decreases as elevation increases above sea level.

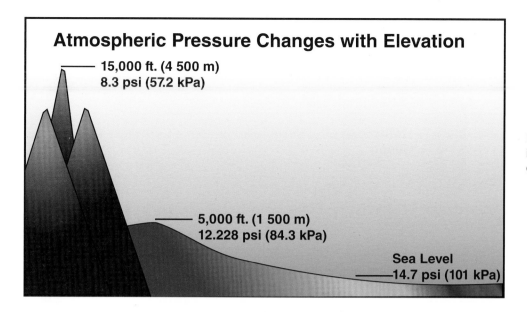

Atmospheric Pressure Changes with Elevation

15,000 ft. (4 500 m)
8.3 psi (57.2 kPa)

5,000 ft. (1 500 m)
12.228 psi (84.3 kPa)

Sea Level
14.7 psi (101 kPa)

Figure 3.13 As the altitude increases, atmospheric pressure decreases.

gauges indicate the psi (kPa) above the existing atmospheric pressure. For example, a gauge reading of 10 psi (69 kPa) at sea level is actually measuring 24.7 psi (170 kPa). Engineers distinguish between a gauge reading and total atmospheric pressure by writing "psig," which is the gauge reading. A notation of "psia" means the absolute or the total atmospheric pressure. A vacuum is any pressure less than the current atmospheric pressure. In common fireground and fire system applications, gauge pressures are almost always used.

Head Pressure

The term **head** in the fire service is another way of expressing pressure. It refers to the pressure created by a column of liquid. Pressure is gained as elevation increases. If a water supply is 100 feet (30 m) above a hydrant discharge opening, it is said to have 100 feet (30 m) of head. To convert this figure to head pressure, divide the number of feet by 2.31. The result is 43.4 psi (300 kPa) at the hydrant **(Figure 3.14, p. 80)**. In metrics, divide the number of meters by 0.1 to get head pressure in kPa.

Elevation affects the pressure or flow of water. Elevation refers to the centerline of the source or the bottom of a static water supply source above or below ground level. The height of a water supply above the discharge orifice is called the **elevation head**. If the nozzle or orifice flowing water is below the source, a pressure gain will occur. Alternatively, if the nozzle or orifice is above the source, then there will be a pressure loss. This result is referred to as *elevation head pressure.*

Static Pressure

Static pressure is defined as stored potential energy available to force water through pipes, fittings, fire hose, and adapters. The term *static* means at rest or without motion. Pressure on water may be produced by an elevated water supply, atmospheric pressure, or a pump. If the water is not moving, the pressure exerted is static. True static pressure is rarely found in a municipal water system because water is always flowing due to customer demand. However, the pressure in a water system before water flows from a hydrant is considered static pressure for testing or fire fighting purposes.

Head — Alternate term for pressure, especially pressure due to elevation. For every 1-foot increase in elevation, 0.434 psi is gained (for every 1-meter increase in elevation, 9.82 kPa is gained). *Also called Head Pressure.*

Elevation Head — Pressure related to the difference in elevation between the water supply and the discharge orifice, commonly expressed in feet.

Static Pressure — (1) Potential energy that is available to force water through pipes and fittings, fire hose, and adapters. (2) Pressure at a given point in a water system when no testing or fire protection water is flowing.

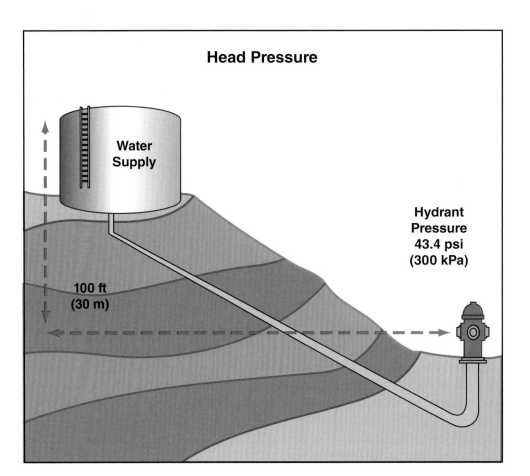

Figure 3.14 The height of the water supply above the fire hydrant creates the head pressure.

Inside figure:

Head Pressure

Water Supply

Hydrant Pressure 43.4 psi (300 kPa)

100 ft (30 m)

Normal Operating Pressure

Normal operating pressure is that pressure found in a water distribution system during normal consumption demands. As soon as water starts to flow through a distribution system, static pressure no longer exists. The demand for water consumption fluctuates continuously, causing water pressure to increase or decrease in the system.

Residual Pressure

Residual pressure is that part of the total available pressure not used to overcome friction loss or gravity while forcing water through pipes, fittings, fire hose, and adapters. Residual means a remainder or that which is left. In a water distribution system, residual pressure varies according to the amount of water flowing from one or more hydrants, water consumption demands, the pipe size, and any other restrictions that are present. For example, during a hydrant fire flow test, residual pressure represents the pressure left in a distribution system within the vicinity of one or more flowing hydrants when the test water is flowing. Residual pressure is measured at the location where a static pressure is taken, not at the point of waterflow.

Flow Pressure (Velocity Pressure)

Flow pressure is the forward velocity pressure while water is flowing. The forward velocity or flow pressure can be measured by using a pitot tube and gauge **(Figure 3.15)**. If the size of the opening is known, a firefighter can use the measurement of flow pressure to calculate the quantity of water flowing in gpm or L/min.

Residual Pressure — Pressure measured at the hydrant to which a pressure gauge is attached while water is flowing from one or more other hydrants during a hydrant flow test. It represents the pressure remaining in the water supply system while the test water is flowing and is that part of the total pressure that is not used to overcome friction or gravity while forcing water through fire hose, pipe, fittings, and adapters.

Flow Pressure — Pressure created by the rate of flow or velocity of water coming from a discharge opening.

Figure 3.15 A pitot tube is used to measure the velocity pressure at the hydrant.

Friction Loss

Friction loss is another concept in which fire and emergency service personnel must be familiar. The fire service definition of friction loss is that part of the total pressure lost while forcing water through pipes, fittings, fire hose, and adapters. In order to effectively assess waterflow for fire protection, friction loss must be understood and taken into account.

Friction loss occurs for a variety of reasons, including the following:

- Movement of water molecules against each other and against the interior surfaces of piping and hose
- Condition, age, or interior surface of pipe or hose
- Couplings, valves, appliances, and fittings
- Sharp bends
- Diameter of pipe or hose
- Improper gasket size

Friction loss can be measured by inserting in-line gauges in a hose or pipe where the pipe or hose diameter and elevation do not change. Friction loss is the difference in the residual pressures between gauges when water is flowing. For example, the difference in the pressure between the nozzle and the pumper is a good example of friction loss where the hose diameter remains constant and the nozzle is at the same elevation as the pumper outlet. The four basic principles that govern friction loss in hose and pipes are described in the following section.

First Principle

If all other conditions are the same, friction loss varies directly with the length of the hose or pipe. Friction loss increases as the length of hose or piping increases. Comparing two pipes that vary only in length, one 100 feet (30 m) and the other 200 feet (60 m), both maintaining a constant flow of 200 gpm (750 L/min) will result in a friction loss of 10 psi (70 kPa) for the 100-foot (30 m) length and 20 psi (140 kPa) for the 200-foot (60 m) length **(Figure 3.16, p. 82)**.

> **Friction Loss** — That part of the total pressure lost as water moves through a hose or piping system; caused by water turbulence and the roughness of interior surfaces of hose or pipe.

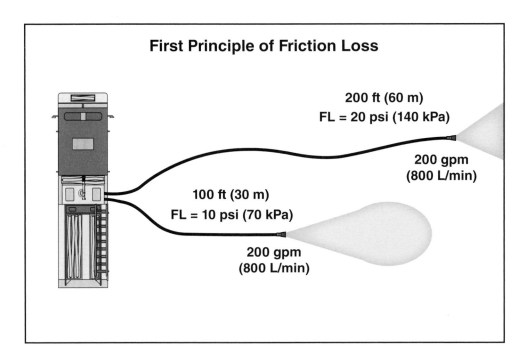

Figure 3.16 In the first principle of friction loss, if all other conditions are the same, friction loss varies directly with the length of the hose or pipe.

First Principle of Friction Loss

200 ft (60 m)
FL = 20 psi (140 kPa)

200 gpm
(800 L/min)

100 ft (30 m)
FL = 10 psi (70 kPa)

200 gpm
(800 L/min)

Second Principle

When hoses or pipes are the same diameter, friction loss varies approximately with the square of the increase in the flow velocity. Friction loss develops much faster than the change in velocity. For example, a pipe flowing 200 gpm (800 L/min) has a friction loss of 3 psi (21 kPa), while doubling the flow to 400 gpm (1 600 L/min) increases the friction loss by about four times to approximately 12 psi (84 kPa). Tripling the original flow to 600 gpm (2 400 L/min) increases the friction loss by about nine times to approximately 27 psi (189 kPa) **(Figure 3.17)**.

Third Principle

For the same discharge, friction loss varies inversely to approximately the fifth power of the diameter of the hose or pipe. This principle illustrates the advantage of a larger diameter hose or pipe. For example, when comparing the friction loss in a pipe that is 2½ inch (65 mm) in diameter with a pipe that is 3 in (77 mm) in diameter with the same flow rate in both pipes, the friction loss in the 3-inch (77 mm) pipe is about 0.4 that of the 2½ in (65 mm) pipe.

Fourth Principle

When hoses or pipes are the same diameter, same length, and have the same flow, friction loss varies inversely with approximately the square of the coefficient of roughness for the pipe or hose. The higher the coefficient of roughness, the less friction loss will be encountered. For example, a 100-foot (30 m), 8-inch diameter (200 mm) pipe flowing 2,750 gpm (10 410 L/min) with a coefficient of roughness of 100 has a friction loss of approximately 8 psi (55 kPa). Assuming the flow, length, and diameter are the same, a pipe with a coefficient of roughness of 120 has a friction loss of approximately 6 psi (40 kPa), and a pipe with a coefficient of roughness of 140 has a friction loss of approximately 4 psi (30 kPa).

Characteristics of hose and piping layouts, such as length and diameter, affect friction loss. Sharp bends or kinks in the hose, as well as elbows and other fittings in piping, can increase friction loss as well. Using large diameter hose and removing sharp bends or kinks in fire hose will reduce the amount of friction loss. In fire fighting operations, any extra hose should be eliminated to reduce excess friction loss.

Public Water (Potable) Supply Systems

Water for human consumption comes from one of two basic sources: a well or a public (municipal) water system. Water supply is essential for business and industry to operate and for adequate fire protection. The objectives of a public water supply system are to provide adequate and reliable services to its customers, which should include meeting fire flow and domestic needs.

Maintaining a continuous or uninterrupted water supply for public demands is a major challenge to many municipalities because of the following conditions:

- Droughts
- Growing demands that cannot be met by the treatment plant
- Lack of adequate storage capacity
- Other communities drawing water from the same supply sources such as a lake or river
- Major commercial fire or wildland/urban interface fire that exhausts the water supply
- Undetected underground leakage of the pipe distribution system
- Contamination
- Aging and deterioration of water distribution network

A municipality must recognize that the quantity of available water needs to be such that maximum daily consumption demands are satisfied at all times, even during periods of drought or after years of community growth. The water delivery system should expand as the municipality increases.

Public water systems provide the methods for supplying water to communities. While water systems are complicated, each must incorporate the following basic components:

- Sources of water supply
- Water processing or treatment facilities
- Means of moving water
- Water storage and distribution systems

Sources of Water Supply

Every water supply system must have a supply source that is both adequate and reliable for the area being served. The primary water supply can be obtained from surface water or groundwater. Although most water systems are supplied

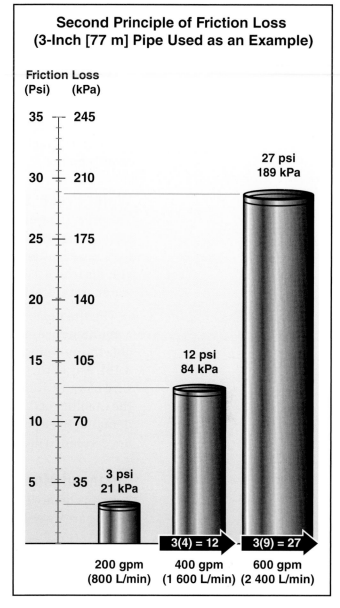

Figure 3.17 Example of the second principle of friction loss.

from only one source, there are instances when both sources are used. Rivers and lakes are two examples of surface water supply. Groundwater supply can be water wells or water-producing springs.

An engineering estimate can determine the amount of water that a facility needs. This estimate is the total amount of water needed for domestic, industrial, and fire fighting use. In cities, the domestic/industrial requirements far exceed those needed for fire protection. However, in some locations, such as industrial facilities, the requirements for fire protection may exceed other requirements.

Water Processing or Treatment Facilities

The processing or treatment facility is the second component of a water supply system **(Figure 3.18)**. The treatment of water is a vital process. Water is treated to remove contaminants that may be detrimental to the health of those individuals who use or drink it. In addition to removing harmful elements from water, water treatment officials may add fluoride or oxygen.

The fire department's main concern regarding treatment facilities is that a maintenance error, natural disaster, loss of power supply, or fire could disable the pumping station or severely hamper the purification process **(Figure 3.19)**. Any of these situations would drastically reduce the volume and pressure of water available for fire fighting operations. Another problem would be the inability of the treatment system to process water fast enough to meet the demand. In either case, fire officials should work with the water purveyor to have a plan to deal with these potential shortfalls.

Means of Moving Water

A means of moving water is the third component of a water supply. The following are three basic means of moving water in a municipal system **(Figure 3.20)**.

Direct Pumping System

This system uses one or more pumps to take water from the primary source and discharge it into the distribution system. Failures in supply lines and pumps can usually be overcome by duplicating these units and providing a secondary power source.

Gravity System

The gravity system uses a primary water source located at a higher elevation than the distribution system. The gravity flow from the higher elevation provides the water pressure. This pressure is usually sufficient only when the primary water source is located at least several hundred feet (meters) higher than the highest point in the water distribution system. The most common examples of gravity systems include a mountain reservoir that supplies water to a city below or a system of elevated tanks or water towers. These tanks provide a reliable water supply, and it is only necessary to keep the tanks full and the valves open to ensure that the water will flow when needed. However, tanks require considerable maintenance and often protection against freezing. They are also frequent targets of vandalism and mischief.

Figure 3.18 Treatment tanks like these are commonly found in water treatment facilities.

Figure 3.19 Failure of pumps like this one can adversely impact fire fighting efforts due to decreased water supply.

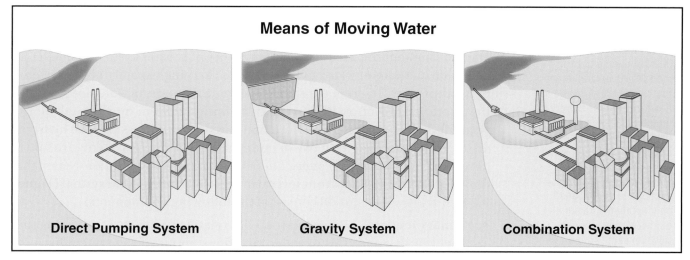

Means of Moving Water

Direct Pumping System   Gravity System   Combination System

Figure 3.20 Three basic ways to move water in a municipal system.

Combination System

This system uses combinations of direct pumping and gravity systems. In most cases, elevated storage tanks supply the gravity flow. These tanks serve as emergency storage and provide adequate pressure through the use of gravity. During periods of low consumption when the system pressure is high, automatic valves open and allow the elevated storage tanks to fill. When the pressure drops during periods of heavy consumption, the storage containers provide extra water by feeding it back into the distribution system. Providing a good combination system involves reliable, duplicated equipment and proper-sized, strategically located storage containers.

Water Storage and Distribution Systems

The storage and distribution systems are the last components of the water supply system. Water storage should be sufficient to provide for domestic and industrial requirements in addition to the demands expected during fire fighting operations. Such storage should also be sufficient to permit making most repairs, alterations, or additions to the system. Location of the storage and the capacity of the mains leading from this storage are also important factors. Water stored in elevated reservoirs can also ensure water supply when the system becomes otherwise inoperative.

Many industries that provide their own private water systems use storage tanks that are available to the fire department **(Figure 3.21)**. Water for fire protection may be available to some communities from storage systems, such as cisterns, that are considered a part of the distribution system. The fire department pumper removes the water from these sources by drafting and provides pressure with its pump.

Water storage becomes an important fire protection feature during times of peak demand. Elevated tanks are used to stabilize or balance the pressure on a water system when demand is high **(Figure 3.22)**. As long as the pumps supplying the system can keep up with the demand, the water in the tanks will not be used. However, once the demand upon a system becomes so great that the pumps cannot keep up, the system pressure will begin to drop. When the pressure decreases to a point where it cannot keep the system full, the tank will begin to add water to the system.

The distribution system receives water from the pumping station and delivers it throughout the area served. The ability of a water system to deliver an adequate quantity of water relies upon the carrying capacity of the system's network of pipes. In order to provide adequate amounts of water for fire fighting operations, friction loss in the distribution system must be taken into account. To reduce the effect of friction loss on water pressure, fire hydrants are typically supplied from two or more directions. These systems are said to have a **circulating feed** or *looped line* **(Figure 3.23)**. A distribution system that provides circulating feed from several mains constitutes a grid system **(Figure 3.24)**. A grid system should consist of the following components:

- **Primary feeders** — Pipes (mains) with relatively widespread spacing that convey large quantities of water to various points of the system for local distribution to the smaller mains.

- **Secondary feeders** — Network of pipes that reinforce the grid within the various loops of the primary feeder system and aid the delivery of the required fire flow at any point.

- **Distributors** — Grid arrangement of mains serving individual fire hydrants and blocks of customers.

- **Circulating feed** — Fire hydrant that receives water from two or more directions. To ensure sufficient water, two or more primary feeders should run from the supply source to the high-risk and industrial districts of the community by separate routes. Similarly, secondary feeders should be arranged in loops as far as possible to give two directions of supply to any point. This practice increases the capacity of the supply at any given location and ensures that a break in a feeder main will not completely cut off the water supply.

Circulating Feed — Water main piping interconnected to form a loop or grid arrangement that allows water to flow from two or more directions when serving a fire hydrant, building water service or dead-end water main. Circulating feeds allow water distribution systems to be designed and installed with smaller diameter pipe that can flow the same amount of water than larger diameter piping served from one direction only.

Figure 3.21 Some industrial sites have water storage capabilities that are made available to the fire department.

Figure 3.22 Elevated tanks are used to balance the pressure on a water system during periods of peak demand.

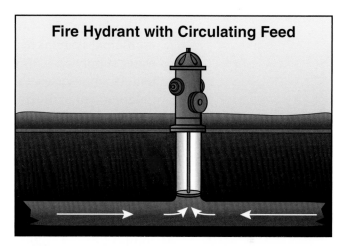

Figure 3.23 Looped fire hydrants receive water from two directions.

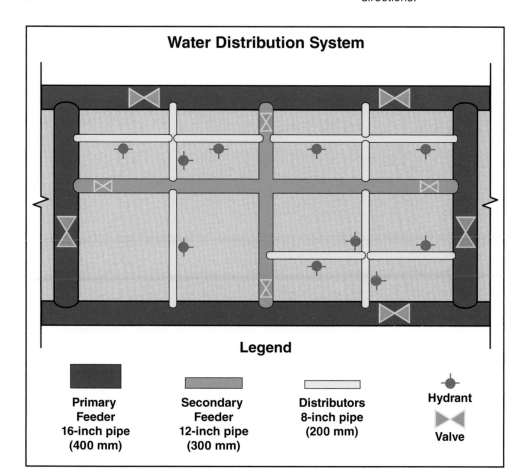

Figure 3.24 A typical grid system of water supply pipes.

Private Water Supply Systems

In addition to the public water supply systems that service most communities, fire and emergency services personnel must also be familiar with the basic principles of any private water supply systems that are within their response jurisdiction. Private water supply systems are most commonly found on large commercial, industrial, or institutional properties **(Figure 3.25)**. These systems may service a large building or a series of buildings on the complex. In general, a private water supply system exists to provide water for fire protection purposes, sanitary purposes, manufacturing processes, or a combination of these uses.

The design of private water supply systems is typically similar to that of the municipal systems described earlier in this chapter. Most commonly, private water supply systems receive their water from a municipal water supply system. In some cases, the private system may have its own water supply source independent of the municipal water distribution system.

Two sources of water supply may occasionally serve properties for fire protection: one from the municipal system and the other from a private source. In many cases, the private source of water for fire protection provides nonpotable water. When this is the case, adequate measures must be taken to prevent

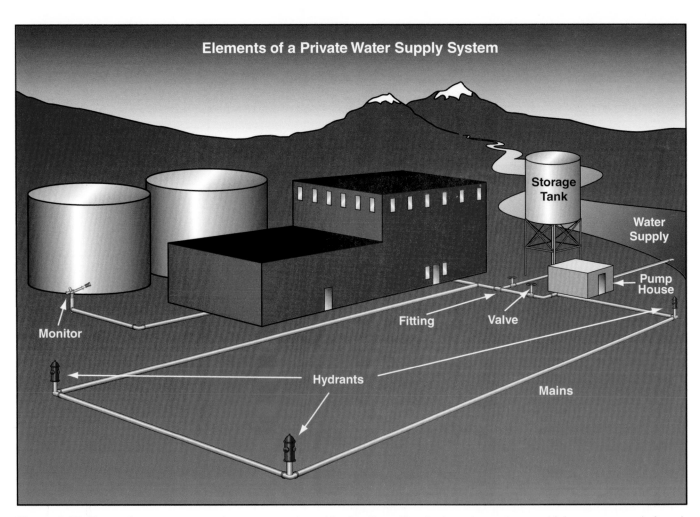

Figure 3.25 Fire and emergency services personnel must be familiar with private water systems, which are commonly found on industrial sites.

contamination caused by the backflow of water into the municipal water supply system. A variety of backflow prevention measures can be employed to avoid this problem. Some jurisdictions do not allow the interconnections of potable and nonpotable water supply systems. This separation means that the protected property is required to maintain two completely separate systems.

Many private water supply systems maintain separate piping for fire protection, including hydrants, and for domestic and industrial uses. Separate systems are cost-prohibitive for most municipal applications but are economically feasible in many private applications. A number of advantages to having separate piping arrangements in a private water supply system include the following:

- The property owner has control over the water supply source.
- Neither system (fire protection or domestic/industrial) is affected by service interruptions to the other system.
- The facility has the ability to isolate systems for inspection, testing, and maintenance.

Fire and emergency services personnel must be familiar with the design and reliability of private water supply systems in their jurisdiction. Large, well-maintained systems may provide a reliable source of water for fire protection purposes. Small capacity, poorly maintained, or otherwise unreliable private water supply systems should not be depended upon to provide all the water necessary for adequate fire suppression operations. Historically, many significant fire losses can be traced, at least in part, to the failure of a private water supply system that was being used by municipal fire departments working the incident. In the event of a fire, disastrous losses may result when electrical service is lost to a property protected by electrically driven fire pumps.

Water Piping Systems

Having a large capacity and reliable supply of water is of little value if the water cannot be delivered in adequate amounts and with adequate pressure to the point of use. This requirement depends heavily upon the capacity and pressure rating of the pumps at the treatment plant and pumping stations, as well as the extent of elevated storage. The piping distribution system is the feature that has the greatest impact. Several variables affect the performance of the distribution system, including:

- Piping materials
- Piping fittings
- Pipe diameter
- Piping arrangements

In addition, water main valves and fire hydrants are important parts of the water piping system, affecting the system's reliability. Their location is of particular importance for hydrants.

Piping Materials

Piping systems are generally made of cast iron, ductile iron, asbestos cement, steel, polyvinyl chloride (PVC) plastic, or concrete. Whenever a water main is installed, it must be the proper type for the soil conditions and pressures to

Figure 3.26 Internal pipe corrosion accumulates on the interior surface of the pipe.

which it will be subjected. When water mains are installed in unstable or corrosive soils, steel or reinforced concrete pipe or approved casing materials may be used to give the strength needed. Some locations that may require extra protection include areas beneath railroad tracks and highways, areas close to heavy industrial machinery, or areas prone to earthquakes.

The internal surface of the pipe should be such that it offers the least potential for friction loss. Some materials have considerably less resistance to waterflow than others. In addition to the internal surface itself, friction loss can be increased by encrustation on the interior surface of the pipe, corrosion, or sedimentation **(Figure 3.26)**. All of these issues can result in a restriction of the pipe size, causing a reduction in the amount of water that can be supplied by the system.

Pipe Fittings

Pipe fittings are generally made of cast iron, malleable iron, steel, or chlorinated polyvinyl chloride (CPVC), depending upon the type of pipe being connected and the installation conditions encountered. Fittings are required to meet the construction standards put forth by ASME and ASTM to ensure they are robust enough to handle the anticipated operating conditions. When determining which type of fittings to use, an individual must evaluate the compatibility of the fitting material and the piping material, as well as consider the need for restraint against movement at changes in direction. All changes in direction of underground piping must either be supported by thrust blocks or constructed as restrained joints to prevent the underground piping from becoming dislodged at the fittings due to the forces exerted by the moving water within the system. Consider the anticipated soil conditions to be encountered to determine the need for exterior coatings on the fittings to protect against corrosion. In general, it is always best to utilize fittings with the greatest bend radius possible to help reduce the amount of friction created as water moves through the fitting, decreasing the amount of pressure lost due to friction. Reducing the number of fittings is also a key component to lessening the overall pressure loss due to friction within a system.

Pipe Diameters

Water distribution piping should be designed to accommodate current and future domestic and fire flow demands. Fire hydrant supply mains should be at least 6 inches (150 mm) in diameter. These mains should be closely gridded by 8-inch (200 mm) or larger mains at intervals of no more than 600 feet (180 m). Since the water systems in many communities were constructed with little attention to fire protection demands, it is common to see 4-inch (100 mm) pipe supplying fire hydrants in the older sections of many cities and towns.

Piping Arrangements

Piping arrangements are the third feature that affects water distribution system performance. Piping can be a loop system, grid system, dead-end main, or a combination of them.

A **loop system** is an arrangement of water mains where the water will be supplied to a given point from more than one direction **(Figure 3.27)**. These mains are also called *circle systems, circulating systems,* or *belt systems.*

A **grid system** is a large piping distribution system that is characterized by multiple water pathways **(Figure 3.28, p. 92)**. A well-designed grid system uses pipe sizes of 8 inches (200 mm) and larger, which is one of the identifying features of a strong distribution system.

A **dead-end main** is a single pipe that extends from a looped or grid system that is supplied from one direction **(Figure 3.29, p. 93)**. This is a less desirable piping arrangement because water is supplied from only one direction and the volume of water available is typically limited.

Loop System — Water main arranged in a complete circuit so that water will be supplied to a given point from more than one direction.

Grid System — Interconnecting system of water mains in a crisscross or rectangular pattern.

Dead-End Main — Water main that is not looped and in which water can flow in only one direction.

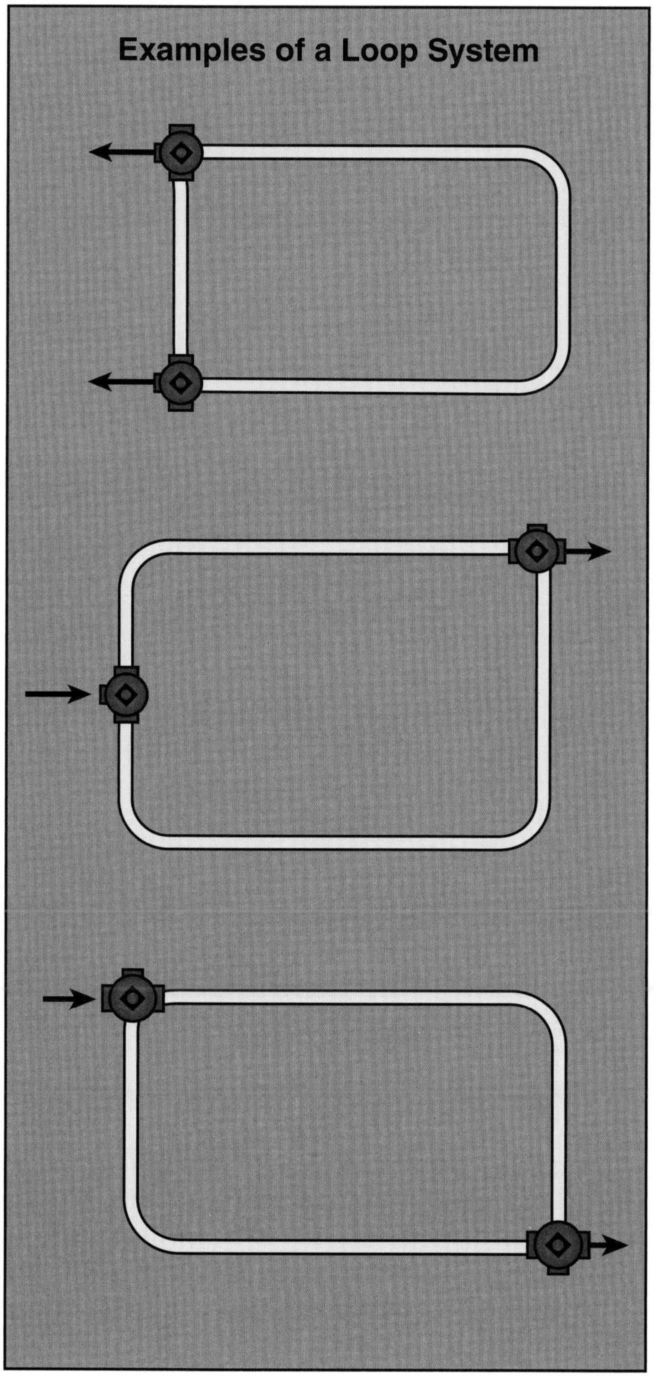

Examples of a Loop System

Figure 3.27 Loop system water mains are also called *circle systems, circulating systems, or belt systems.*

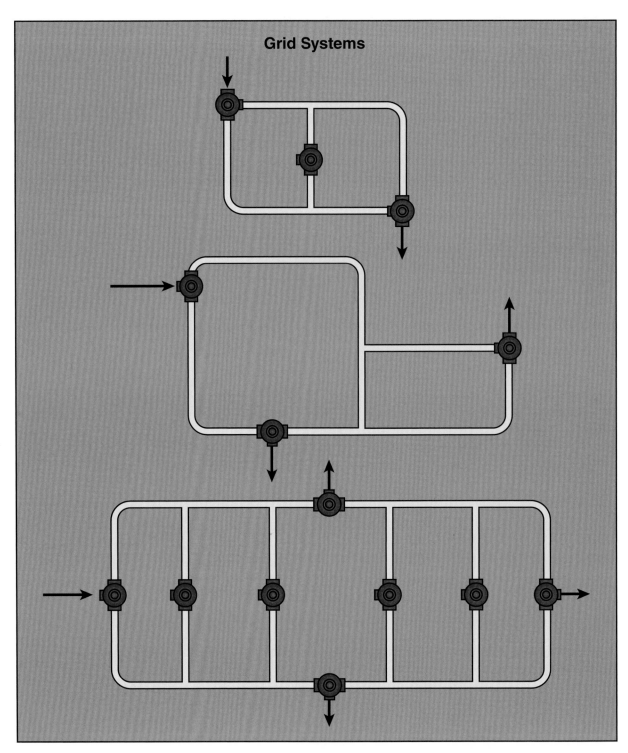

Figure 3.28 A grid system is a redundant distribution system with multiple water pathways.

Water Main Valves

Valves are placed on water distribution systems to provide the ability to isolate portions of the system. This is a particularly important reliability feature. When a water main breaks, strategically closing valves necessitates only a small portion of the system being placed out of service while repairs are made **(Figure 3.30)**. These are typically underground valves that require a special socket wrench on a reach rod, often called a *valve key*, to operate **(Figure 3.31)**.

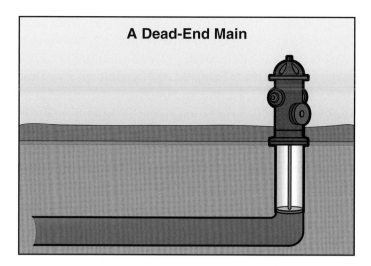

A Dead-End Main

Figure 3.29 This hydrant is served by a dead-end main.

Figure 3.30 Failure of a large water main requires that the break be isolated by closing valves so that repairs can be made. *Courtesy of Tarpon Springs (FL) Fire Rescue.*

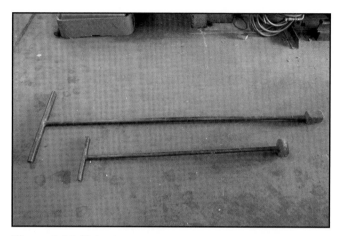

Figure 3.31 Valve keys vary in length and long keys may be needed to reach deep valves.

Ideally, these valves should be located on each branch at the intersection of the mains and no more than 500 feet (150 m) apart in high-value areas. A valve should be on every branch feeding a hydrant so that each hydrant can be isolated during repair or replacement.

Since these valves are underground, access is only possible through metal valve plates located at ground level **(Figure 3.32, p. 94)**. Unfortunately, dirt and pavement often cover these access plates, leaving valves inaccessible and their locations hidden. A good maintenance program will require accurate water maps, indicating valve locations and yearly valve operations to make sure that they are working properly and are completely open **(Figure 3.33, p. 94)**.

Valves for water systems are broadly divided into indicating and non-indicating types. An **indicating valve** visually shows the position of the valve: fully open, fully closed, or partially closed. Valves in private fire protection systems are required to be an indicating type unless specifically approved by

Indicating Valve — Water main valve that visually shows the open or closed status of the valve.

the AHJ per NFPA® 24. Two common indicator valves are the **post indicator valve (PIV)** and the outside stem and yoke (OS&Y) valve **(Figures 3.34a and b)**. The PIV is a hollow metal post that sits on top of the valve housing. The valve stem inside this post has the words "OPEN" and "SHUT" visible on it so that the position of the valve is shown **(Figures 3.35a and b)**. The OS&Y valve has a yoke on the outside of the structure with a threaded stem that opens or closes the gate inside the valve. The threaded portion of the stem is out of the yoke when the valve is open and inside the yoke when the valve is closed **(Figure 3.36, p. 96)**. These valves are commonly used on sprinkler systems but may be found in some water distribution system applications.

Nonindicating valves in a water distribution system are normally buried or located in utility openings. These are the most common types of valves used on most public water distribution systems. If a buried valve is properly installed, the valve can be operated aboveground through a valve box or road box.

Control valves in water distribution systems may be either gate valves or butterfly valves. Gate valves may have a rising stem or a non-rising stem. The rising stem type is similar to the OS&Y valve. On the nonrising stem type, the gate either rises or lowers to control the waterflow when the valve nut is turned by the valve key or wrench. Nonrising stem gate valves should be marked by a number indicating the number of turns necessary to completely close the valve. If a valve resists turning after fewer than the indicated number of turns, it usually means that debris or other obstructions are in the valve.

Butterfly valves are tight closing and usually have a rubber or a rubber-composition seat that is bonded to the valve body **(Figure 3.37, p. 96)**. The valve disk rotates 90 degrees from the fully open to the tight-shut position. The nonindicating butterfly types also require a valve key. Its principles of operation provide satisfactory water control after long periods of inactivity.

A high level of friction loss will result if valves are partially closed. When valves are closed or partially closed, the condition may be unnoticeable during ordinary domestic flows of water. As a result, the impairment will be unknown until a fire occurs or detailed inspections and fire flow tests are done. A fire department will experience difficulty in obtaining water in areas where there are closed or partially closed valves in the distribution system. When valves are partially closed, pump operators may notice a normal static pressure and a severe (abnormal) pressure drop in residual pressure when attempting to flow supply lines or large hose streams.

Figure 3.32 A typical valve access plate.

Figure 3.33 Yearly valve operations ensure that the valve is operating properly and that it is fully open.

Figure 3.34a A typical post indicator valve (PIV).

Figure 3.34b A series of outside stem and yoke (OS&Y) valves.

Figure 3.35a A PIV in the open position reads OPEN in its window.

Figure 3.35b A PIV in the closed position reads SHUT in its window.

Figure 3.36 In OS&Y valves, the threaded portion of the stem is out of the yoke when the valve is open.

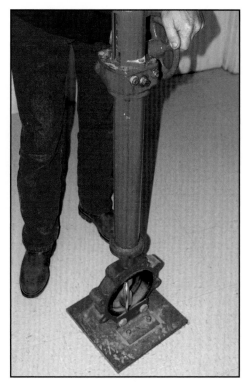

Figure 3.37 A typical butterfly valve.

Dry-Barrel Hydrant— Operating valve underground at the base of the hydrant rather than in the aboveground portion of the hydrant barrel. The valve actually is in the hydrant barrel and not actually at the water main. When operating properly, there is no water in the barrel of the hydrant when it is not in use. These hydrants should be used in areas where freezing could occur.

Fire Hydrants

Fire hydrants are the primary means by which fire and emergency services personnel obtain water for fire suppression. Hydrants are available in two basic types: dry-barrel and wet-barrel. The **dry-barrel** type is the most common type of hydrant used in the United States. As the name implies, there is no water inside the hydrant until the hydrant is turned on **(Figure 3.38)**. This hydrant is typically used in areas where freezing temperatures are expected. The valves in the hydrant keep water out of the barrel until it is opened, and drains located at the bottom of the hydrant allow the water to drain out when the hydrant is closed. When the hydrant is fully opened, the drain holes are closed. If the hydrant is only partially opened, the drain holes will be partially open, permitting a pressurized stream of water to be discharged out of the drains beneath the ground. This drainage will erode the area at the hydrant's base. For this reason, dry-barrel hydrants should always be completely open when in use. Dry-barrel hydrants are easily recognized by the existence of the operating nut on the bonnet or top of the hydrant.

A common variety of dry-barrel hydrant has two 2½-inch (65 mm) outlets and one 4½-inch (115 mm) outlet **(Figure 3.39)**. The larger outlet is often called a pumper or steamer outlet because it is common for the fire department to attach its soft suction line to this outlet to feed the pumping apparatus. It is also common to see dry-barrel hydrants with only two 2½-inch (65 mm) outlets and no pumper outlet. These outlets limit the hydrant's capacity. Therefore, hydrants of this type should not be used in industrial and commercial areas. Several other combinations of outlets may be encountered, such as hydrants having only one or two pumper outlets and no smaller outlets. However, the first two hydrants described are by far the most common.

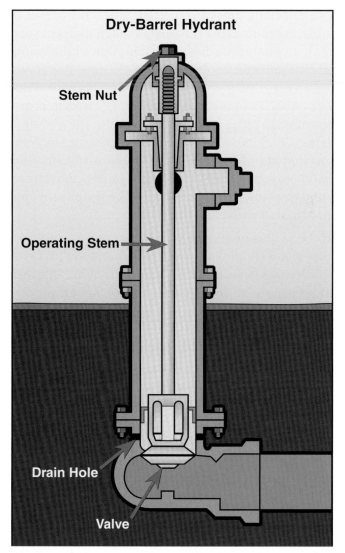

Dry-Barrel Hydrant

Stem Nut

Operating Stem

Drain Hole

Valve

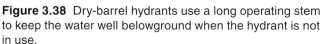

Figure 3.38 Dry-barrel hydrants use a long operating stem to keep the water well belowground when the hydrant is not in use.

Figure 3.39 A variety of a dry-barrel hydrant.

Wet-barrel hydrants may be found where freezing temperatures are un-expected, such as parts of Arizona or southern Florida. Water is inside the hydrant at all times and is subject to freezing in cold climates **(Figure 3.40, p. 98)**. The operating nuts for turning on the hydrant extend through the side of the barrel, and there must be an operating nut for each outlet on the hydrant.

Whether they are wet-barrel or dry-barrel, hydrants are attached to the water main by a short piece of pipe called a *branch*. This branch should be at least 6 inches (150 mm) in diameter. Branches that are 4 inches (100 mm) in diameter are common on older systems, but this size limits the capacity of the hydrants. Each branch should also have a valve so that the hydrant can be isolated for repair or replacement without shutting down the water main to which it is attached.

Grass, shrubs, and other obstructions, such as retaining walls and fences, should be kept away from hydrants so that they remain visible and acces-sible. Hydrants should be flushed every year at a minimum to make sure that they work and to clean sedimentation from the pipes. It is also a good

Wet-Barrel Hydrant — Fire hydrant that has water all the way up to the discharge outlets. The hydrant may have separate valves for each discharge or one valve for all the discharges. This type of hydrant is only used in areas where there is no danger of freezing weather conditions.

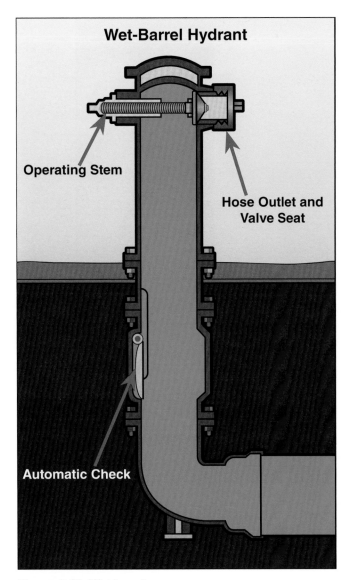

Figure 3.40 Wet barrel hydrants have water right up to the discharge outlets when not in use. This type of hydrant is used in locations where there is no risk of freezing.

idea to annually measure the flow at each hydrant if possible. By comparing the flows at each hydrant year by year, any deterioration or obstruction of the water supply system can be determined. Testing fire flows allows for identification of varying pressures within the water system. In many communities, fire hydrants are color-coded to reflect the flow capacity of the hydrant. Refer to NFPA® 291, *Recommended Practice for Fire Flow Testing and Marking of Hydrants*, for more information. Fire and emergency services organizations must have a strong working relationship with the water provider to coordinate testing and maintenance.

NOTE: Additional information on testing fire flows and fire water supply sources can be found in FPP's *Fire Service Hydraulics and Water Supply* manual. This text provides a more in-depth discussion of these topics.

Hydrant Locations

In order to be effective, fire hydrants are installed at locations with spacing consideration for convenient use by fire departments. They must also be located to meet the needed fire flows for the buildings being protected. Proper hydrant distribution and location are an important feature of an accessible water supply system. The pumper outlet should be readily visible and positioned in the most advantageous location for connection to fire apparatus **(Figures 3.41a and b)**. In addition, the hydrant should be set high enough that the hydrant wrench can be turned a full revolution without coming in contact with the ground when removing the cap from the lowest outlet (usually the pumper outlet).

The Insurance Services Office, Inc. (ISO), in its Fire Suppression Rating Schedule (FSRS), provides a method for locating fire hydrants within a community. This method is based upon its grading schedule criteria for protection classifications. The ISO evaluation procedure looks at a community's population, the number of hydrants in the system, and the property types. Communities are given credit for those hydrants that are within 1,000 feet (300 m) of the property to be protected. In addition, hydrants must deliver a minimum of 250 gpm (1 000 L/min) at 20 psi (140 kPa) residual pressure for a duration of two hours. In more congested areas and high fire-risk areas, it is recommended that hydrants be located 300 feet (90 m) apart. One hydrant should be located at every street intersection, one in the middle of long blocks, and one near the end of a long dead-end street.

The International Fire Code®, 2012 edition, Section 507, provides the following requirements for hydrant locations:

- Where a portion of a facility or building is more than 400 feet (120 m) from a hydrant, on-site fire hydrants and mains shall be provided where required by the fire code official.

Figures 3.41a From this direction, the hydrant is not visible to responding emergency responders.

Figure 3.41b The positioning of this hydrant would make removal of the hydrant caps with a hydrant wrench difficult.

- For certain occupancies (R-3 and Group U), the distance requirement is 600 feet (180 m).

- If the building is equipped throughout with an automatic fire sprinkler system, the distance requirement is 600 feet (180 m).

While different sources provide information on the recommended spacing of hydrants, it is ultimately up to the authority having jurisdiction (AHJ) to make these decisions.

Chapter Summary

Water continues to be the most readily available and plentiful fire extinguishing agent. In addition to knowing that water is available for fire fighting operations, understanding how water extinguishes fire and the advantages and disadvantages of its use is important. In any kind of distribution system, water must be available at sufficient quantities and adequate pressures to be effective. Personnel who must inspect or use these systems need to know how water is distributed as well as any physical and design factors that can positively or negatively affect its availability. The work that water purveyors and fire and emergency services personnel do to keep these systems operating effectively is critical because their purpose is to save lives and property.

Review Questions

1. What are the advantages and disadvantages of using water as an extinguishing agent?

2. What are the basic principles of pressure?

3. Identify four different types of pressure as they relate to water.

4. What are three basic principles of friction loss?

5. What four main components make up a grid system in a public water supply system?

6. What are two advantages to having separate piping arrangements in private water supply systems?

7. Identify and describe two variables that affect the performance of a water distribution system.

8. What are the two main types of valves in water supply systems?

9. Identify and describe the two basic types of fire hydrants.

10. What are two sources that provide methods and recommendations for locating fire hydrants?

◼ Chapter Contents

Key Terms

FESHE Outcomes

Fire and Emergency Services Higher Education (FESHE) Outcomes: Fire Protection Systems

1. Explain the benefits of fire protections systems in various types of structures.
4. Identify the different types and components of sprinkler, standpipe, and foam systems.
10. Discuss the appropriate application of fire protection systems.

Water-Based Fire Suppression Systems

Learning Objectives

After reading this chapter, students will be able to:

1. Describe the benefits of automatic sprinkler systems and reasons for their installation.

2. Describe the design principles for residential and commercial sprinkler systems.

3. Identify the components of an automatic sprinkler system.

4. Describe the most important aspects of sprinkler systems.

5. Identify types of specialty sprinklers and nozzles.

6. Summarize the four most common types of sprinkler systems.

7. Summarize the use and design of water-mist fire suppression systems.

8. Describe the basic principles of foam and foam generation.

9. Identify types of foam fire suppression systems.

10. Identify types of foam nozzles and sprinklers.

11. Summarize considerations for the testing and inspection of sprinkler systems.

Chapter 4
Water-Based
Fire Suppression Systems

Case History

In 2013, a major fire at the Kiss nightclub in Santa Maria, Rio Grande do Sul, Brazil, resulted in the deaths of 242 people and injured around 200 more. According to investigators, the fire was caused when a band ignited a pyrotechnic device, causing the acoustical foam on the ceiling to ignite. Overcrowding and a lack of adequate emergency exits contributed to the large loss of life. The fire had several similarities to the Station nightclub fire in Rhode Island in 2003 that killed 100 people, including that neither club had a fire sprinkler system.

Automatic sprinkler systems can be highly effective components of any building's fire protection plan. According to the U.S. Fire Administration (USFA), the presence of properly designed, installed, and maintained sprinkler systems reduces the chances of death from fire by two-thirds and property losses by 50 percent or more.

The NFPA® wrote the first standard for the design, installation, testing, and inspection of automatic sprinkler systems in 1896. That standard, NFPA® 13, *Standard for the Installation of Sprinkler Systems*, has been in continuous publication since that date. Two additional NFPA® standards regulate the installation of automatic sprinkler systems: NFPA® 13D, *Standard for the Installation of Sprinkler Systems in One- and Two-Family Dwellings and Manufactured Homes*, and NFPA® 13R, *Standard for the Installation of Sprinkler Systems in Low-Rise Residential Occupancies*. These documents have been revised numerous times to reflect new technology, knowledge, and actual loss experiences.

Today, automatic sprinkler systems are unsurpassed in fire protection. According to the NFPA® Report, *U.S. Experience with Sprinklers*, **sprinklers** were effective 96 percent of the time when operated. Failure to control the fires in sprinkler-equipped buildings was reportedly due to the following reasons:

- Closed valves or partially closed control valves
- Improper maintenance
- Inadequate or inoperative water supply

Automatic Sprinkler System — System of water pipes, discharge nozzles, and control valves designed to activate during fires by automatically discharging enough water to control or extinguish a fire. *Also known as* Sprinkler System.

Sprinkler — Waterflow discharge device in a sprinkler system; consists of a threaded intake nipple, a discharge orifice, a heat-actuated plug, and a deflector that creates an effective fire stream pattern that is suitable for fire control.

- Incorrect design for the current hazard
- Obstructions
- Partial protection

The purpose of this chapter is to address the various components and types of automatic sprinkler systems. Aspects of sprinklers and specialty sprinklers are also addressed. Procedures for the inspection and testing of sprinkler systems are highlighted as well. The information in this chapter is intended to be informative and descriptive, but it should not be used as the authority over local codes, ordinances, and standards.

While there are a few commonly used agents in automatic fire sprinkler systems, the most common agent is water. This chapter addresses water-based automatic sprinkler systems and provides an overview of foam systems. Chapter 7, Non-Water-Based Fire Suppression Systems, specifically addresses suppression systems using other agents.

Purpose of Automatic Sprinkler Systems

Modern automatic sprinkler system technology is both highly effective and reliable when properly designed, installed, and maintained. Sprinkler systems are installed in all types of structures, and many jurisdictions require the installation of automatic sprinkler systems in one- and two-family dwellings In fact, several U.S. communities have been requiring sprinkler systems in dwellings for several decades.

Fires that involve large unsprinklered properties pose a threat to the community and place an undue burden on fire fighting resources. Reasons that building owners invest in and install automatic sprinklers include the following:

- Code requirements or by variance
- Potential incentives or reductions in insurance rates **(Figure 4.1)**
- Insurance requirements
- General protection of life and property from fire
- Building design flexibility
- Inherent risk

Model building and fire codes require the installation of automatic sprinkler systems. The reasons for mandating sprinklers in buildings arise from a need to protect the community or to protect the occupants in individual buildings

Figure 4.1 Potential incentives, such as lower insurance rates, may encourage homeowners to install automatic sprinklers in their homes.

of high-occupancy design **(Figure 4.2)**. Model codes require the installation of automatic sprinklers in buildings based upon their occupancy, construction type, and size. Some types of occupancies must always have automatic sprinklers. Other occupancies must have sprinklers when the building exceeds a given size established in a building code. Good risk management practices suggest that automatic sprinklers be installed to protect the substantial investment in the structure and business, even if the code does not require it.

Automatic sprinkler systems are also important to the life safety of firefighters. Controlling a fire in its earlier stages allows for a more **tenable atmosphere** for fire fighting operations. This practice will increase the level of safety within the environment for firefighters to conduct suppression activities. Therefore, fire fighting personnel must understand how automatic sprinkler systems operate so that operations can be conducted more efficiently in buildings protected by these systems **(Figure 4.3)**.

Tenable Atmosphere — Capable of maintaining human life.

The ideal fire control system should be simple, reliable, and automatic. It should use a readily available and inexpensive extinguishing agent and discharge the extinguishing agent directly on the fire while in its incipient stage. An automatic sprinkler system most closely meets these requirements.

Design Principles (Residential/Commercial Systems)

With few exceptions, automatic sprinkler systems within commercial properties are designed to protect all occupied areas and concealed spaces within a building and intended for property protection. Sprinklers are provided within prescribed spacing guidelines as required for the hazard to be protected in accordance with NFPA® 13, *Standard for the Installation of Sprinkler Systems.* The higher the hazard to be protected, the greater the density (gpm/ft² [L/

Figure 4.2 Building codes require multistory occupancies to incorporate automatic sprinkler systems to protect the occupants in individual buildings.

Figure 4.3 All components of a sprinkler system serve specific purposes. First responders should understand how these systems operate.

(m².min)]) required and smaller area to be protected. For example, an ordinary group two hazard, which has a moderate amount of combustibles and storage, has a more restrictive (greater) design density when compared to a light hazard occupancy which typically has a lower amount of combustibles; the light hazard occupancy's sprinkler system would be designed based on a lower design density.

An increase in density and reduction in area would usually require a higher flow rate of water when compared to less hazardous design classifications. In addition to the hazard classification dictating the minimum density and maximum sprinkler area, the hazard classification controls the overall size of the sprinkler zone. Light hazard occupancies are permitted to be protected with a single sprinkler system covering a floor area up to 52,000 square feet (4 800 m²). However, a sprinkler system protecting a high-piled storage area would only be permitted to cover a floor area up to 40,000 square feet (3 700 m²).

Residential sprinkler systems designed for one- and two-family dwellings and multifamily residential occupancies (up to four stories in height) are primary life safety systems intended to suppress fires in the fire growth stage prior to flashover. Therefore, all residential sprinklers are required to have a thermal element or bulb, which is listed as "quick response." Residential sprinklers are also required to have a higher wall-wetting capability when compared to a standard sprinkler.

Residential sprinkler systems are designed to provide the minimum required waterflow and pressure to a specific number of sprinklers within the largest room. Commercial sprinkler systems, meanwhile, are designed to provide the minimum required design density to all sprinklers within a required design area. As a life safety system, residential sprinklers are not required within normally unoccupied areas (attics and garages) and smaller areas (bathrooms and closets) as permitted by NFPA® 13D, *Standard for the Installation of Sprinkler Systems in One and Two Family Dwellings and Manufactured Homes*, and NFPA® 13R, *Standard for the Installation of Sprinkler Systems in Low-Rise Residential Occupancies*.

Residential Systems

A residential sprinkler system is an automatic sprinkler system that is specifically designed to enhance the survivability of individuals who are in the room of fire origin. The primary purpose of these types of systems is to reduce injury or loss of life due to a fire. An added benefit is the potential reduction of property damage. Residential systems are expected to prevent flashover in the room of fire origin and improve the chance for the occupants to escape or be evacuated.

 Residential systems designed for one- and two-family dwellings are smaller and more economical than those for commercial occupancies. The water supply source is generally the same as the domestic water supply. This approach works because the water supply requirements are substantially less for residential systems. A minimum of a 10-minute supply of stored water is required for systems designed in accordance with NFPA® 13D. If structures are less than 2,000 square feet (185 m²) in area and no more than one story in height, the minimum quantity of water is based on the flow rate for two sprinklers multiplied by seven minutes of operation. In residential structures, such as hotels and motels that are four stories or less, NFPA® 13R requires a minimum of a thirty-minute supply.

To make sprinkler systems useful in residential application, a few changes in design, operation, water supply, and flow requirements must be made. These changes decrease the cost of the system while enhancing its effectiveness in protecting life and property. Some of these changes include the following:

- Modification of sprinkler design and the development of quick response residential sprinklers
- Minimum flow requirements of 18 gpm (70 L/min) from an individual sprinkler for residential protection
- Alarms that are simpler and better designed for residential applications

A major difference between residential sprinklers and standard sprinklers is a higher wall wetting requirement for residential sprinklers to prevent flashover. Residential sprinklers are life safety sprinklers, whereas commercial-use sprinklers are intended for property protection. Residential sprinklers operate more quickly than standard sprinklers. By redesigning the fusible link, even at 165°F (74°C), the sprinkler can be made to operate before conditions in the room become so untenable that occupants cannot survive.

Sprinkler coverage in residential systems is not as extensive as in standard commercial systems. Sprinklers can be omitted from areas such as garages, carports, closets, and small bathrooms. However, local codes may be amended to include some of these traditionally exempted areas. Residential sprinklers are designed to discharge water higher on the walls of a room to prevent a fire from traveling above the spray, which might occur with burning drapes or in preflashover conditions.

These systems are installed with a variety of piping materials not generally found in commercial installations. In addition, piping methods, such as multipurpose or combination domestic and sprinkler water lines, are sometimes used to supply both the domestic needs of a residence and the fire sprinklers **(Figure 4.4)**.

To be of value, a residential sprinkler system must continually be in service. As with a standard system, the inadvertent or deliberate closing of valves renders the system useless. Therefore, using one valve to control both the sprinklers and the water service for the residence eliminates the possibility of the supply valve being turned off. The sprinklers cannot be turned off without the household water supply being turned off as well.

The water supply for residential sprinklers may be taken from several sources. These sources can include a connection to the public water system, an on-site pressure tank,

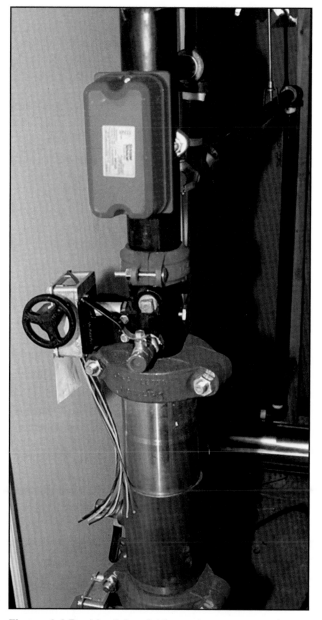

Figure 4.4 Residential sprinkler systems may employ features not found in industrial applications because residential fires tend to include hazards higher along the walls.

Figure 4.5 High piled storage creates the potential for greater loss in a fire.

or a storage tank with an automatic pump. A connection to the public water supply is reliable and will usually provide adequate volume. Public water systems may not service rural homes, which would require the use of a pressure tank or tank with an automatic pump. Some homes found in urban areas may also need a fire pump and/or tank based on hydraulic calculations.

Storage Facility Systems

Storage facilities present a particular set of fire risk issues depending on the height and types of commodities being stored. If storage space is used efficiently, the burning rate and fire spread is increased should a fire occur. The higher loss potential can be attributed to large undivided areas, large concentrations of materials/commodities in a single fire area, and high stockpiling with narrow aisles **(Figure 4.5)**. To protect these areas effectively, firefighters must have knowledge of the sprinkler system and the storage methods associated with the warehouse. NFPA® standards outline methods for providing fire protection for storage facilities based upon commodity classification, the storage arrangement, and height.

Identifying the commodities that are stored and the method of their storage are essential tasks for analyzing the type of sprinkler protection needed for high-storage areas. There are several major categories of commodity classification for storage facilities. These classifications reflect the burning rate and heat release rate of the commodity as well as the effect that water has on the commodity. Commodity classifications are as follows:

- **Class I** — Generally noncombustible and stored on wood pallets in ordinary packaging. Class I commodities can be packaged in corrugated cardboard or stretch-wrapped as a unit load.

- **Class II** — Noncombustible commodities but packaged in wooden crates or multilayered cardboard cartons.

- **Class III** — Combustible materials, such as wood, paper, or certain plastics, regardless of packaging.

- **Class IV** — Class I, II, or III products that contain limited amounts of Group A plastics. Plastics present a special fire-control problem because they produce more heat per unit of weight than ordinary combustibles. Plastics are divided into three groups based upon their heat per unit of weight **(Table 4.1)**.

The four commodity classifications described above will cover most typical storage occupancies. NFPA® 13 also includes specific requirements for other high-challenge commodities, such as rubber tires, roll paper, baled cotton, and items composed primarily of plastics. Other NFPA® standards contain sprinkler system requirements for other unique hazards. For example, NFPA® 30, *Flammable and Combustible Liquids Code*, and NFPA® 30B, *Code for the*

Manufacture and Storage of Aerosol Products, contain requirements for sprinkler systems protecting these specific hazards.

The classification is based upon the most severe hazard in the storage area. Miscellaneous types of storage have commodities of several classifications. It is better to assign a higher classification than originally contemplated because of the possibility of higher-classed commodities being introduced later. This approach might include a small increase in cost at installation but is typically cheaper than upgrading the system once it has been installed. For more information on commodity classifications, refer to NFPA® 13.

Table 4.1 Types of Plastics		
Fastest Burning **Group A**	Moderate Burning **Group B**	Slowest Burning **Group C**
Acrylic	Cellulose Acetate	Melamine
Polycarbonate	Ethyl Cellulose	Phenolic
Polyethylene	Nylon	Polyvinyl Fluoride
Polystyrene	Silicone Rubber	Urea Formaldehyde

Warehouse protection often involves the use of special sprinklers. These sprinklers may be the following:

- High temperature sprinklers
- Sprinklers with large orifices
- Early Suppression Fast Response Sprinklers (ESFR)
- Control Mode Specific Application (CMSA) sprinklers, formerly called *Large Drop sprinklers*

If rack storage is involved, in-rack sprinklers might also be present.

Pipe Schedule Systems

An early method of establishing pipe diameters uses the number of sprinklers within a design area installed on a pipe segment to determine the minimum diameter permitted. For example, a light hazard occupancy having steel sprinkler pipe would permit:

- Two sprinklers to be fed from 1-inch (25 mm) diameter pipe
- A combined total of three sprinklers to be fed from 1¼-inch (35 mm) diameter pipe
- A combined total of five sprinklers to be fed from 1½-inch (38 mm) diameter pipe
- A combined total of 10 sprinklers to be fed from 1½-inch (38 mm) diameter pipe

This method is still permitted when modifying an existing, approved pipe schedule system and in very limited other circumstances as permitted by NFPA® 13, *Standard for the Installation of Sprinkler Systems*. Special attention should be given to these systems to ensure that the minimum required residual pressure and flow are available as required by NFPA® 13. These systems may be decades old and supplied by public/private water sources that may have lost the initial water pressure or capacity to move large volumes of water.

Hydraulically Calculated Systems

Hydraulically calculated systems are a method of determining sprinkler pipe diameters by calculating the required waterflow (density) and friction loss generated by feed mains supplying pipe segments within a given design area.

For example, a sprinkler protecting an area of 130 feet² (12 m²) having a required density of 0.2 gpm/ft² (8 L/[m².min]) would flow 26 gpm (105 L/min). Where the sprinkler has a K-factor of 8.0, the minimum required pressure

$$\left\{ P = \frac{4.53 \, x \, Q^{1.85}}{C^{1.85} \, x \, D^{4.87}} \right\}$$

provided to the sprinkler would be 10.56 psi (70 kPa) (P = (Q/K)²). Working from the most remote sprinkler within the design area back to the source, the end sprinkler (#1) would require 26 gpm (105 L/min) at 10.56 psi (70 kPa). The next sprinkler 12-feet (3.5 m) upstream (#2), also on 1-inch (25 mm) schedule 40 pipe, would also require 26 gpm (105 L/min) plus the friction loss *(see formula to the left)* associated with the flow to support sprinkler #1. The pressure required at sprinkler #2 would be 10.84 psi (75 kPa), which would discharge 26.34 gpm (105 L/min) (Q = K√P). This process is repeated for all pipe segments flowing sprinklers within the design area and mains serving those sprinklers within the design area.

NOTE: Similar to fireground hydraulics, all flow points are added together. Whereas only the friction loss associated with pipe segments and fittings as well as elevation losses/gains are added to the initial required sprinkler pressure. Much like the passing of pipe schedule systems, hand calculating sprinkler systems are quickly becoming obsolete with the availability of personal computers and calculation software that makes the process more efficient and less time consuming.

Components of Automatic Sprinkler Systems

An automatic sprinkler system is an integrated system of pipes, sprinklers, and control valves designed to activate during fires by automatically discharging enough water to control or extinguish the fire **(Figure 4.6)**. A series of pipes supplies water in these systems to the sprinklers.

Sprinklers in wet, dry, and preaction type sprinkler systems are normally closed. They are kept closed by **thermal elements**. These thermal elements allow the sprinkler to open automatically when heated by a fire. Sprinklers in deluge-type systems are normally open. Water is kept from discharging through the sprinklers by a deluge valve that opens when actuated by a signal from some other means of detection. More regarding each type of sprinkler system is explained later in this chapter. An automatic sprinkler system typically consists of the following components:

- Suitable water supply
- Distribution piping
- Valves

- Sprinklers
- Fire Department Connection (FDC)
- Pipe supports

Thermal Element — Device used in sprinklers and some fire detection equipment that is designed to activate when temperatures reach a predetermined level.

Figure 4.6 Automatic sprinkler system designs are unique to each occupancy. *Courtesy of the Sand Springs (OK) Fire Department.*

Sprinklers are designed to provide protection in a wide variety of situations. Although they may be fairly sophisticated in actual application, the fundamental concept is still simple. It is the simplicity of the automatic sprinkler system that gives rise to its greatest virtue — reliability. The sections that follow describe the individual components of an automatic sprinkler system.

Suitable Water Supply

Every automatic sprinkler system must have a water supply of adequate volume, pressure, and reliability. Minimum waterflow requirements for the system are based upon the hazard being protected, the occupancy classification, and the fuel-loading conditions. The water supply system must be able to deliver the required volume and pressure of water to the highest or most remote sprinkler in a structure while maintaining a minimum residual, or remaining, pressure in the system. Systems must have a primary water supply and may be required to have a secondary water supply.

The primary water supply may come from a public or private source. A private water supply will be necessary if no public supply is available **(Figure 4.7)**. Private water supplies may originate from impounded water sources, such as on-site ponds, reservoirs, wells, and storage tanks

Storage tanks may be used as secondary sources. In some circumstances, such as residential systems, they may also be the primary source. Secondary sources of water supply also include large static water sources. For more information on water supply, refer to Chapter 3, Water Supply Systems.

Distribution Piping

Four basic pipe materials are used in sprinkler systems, including the following:

- Black steel
- Galvanized
- Copper
- Plastic

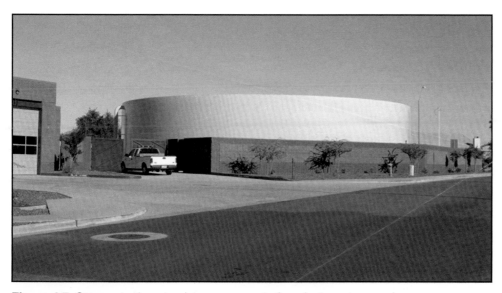

Figure 4.7 Storage tanks are a key component of a private water supply.

The various NFPA® sprinkler system standards specify the types of piping used in a sprinkler system. Black steel piping, also known as *ferrous metal*, is the most common type in use **(Figure 4.8)**. Black steel piping has a long life expectancy but will eventually corrode. In addition, it is very heavy. This type of pipe is typically joined by threading, grooves, flanges, or welding.

Galvanized steel pipe is produced by applying a zinc coating to steel pipe. This protective coating may inhibit the corrosion of the pipe. Galvanized pipe is often required for dry and preaction sprinkler systems.

Copper tubing has been approved for sprinkler systems since the early 1960s. It is joined together by soldering or brazing. Copper tubing is highly resistant to corrosion, has low friction loss, is lighter in weight than steel pipe, and is neat in appearance where exposed. It is rather expensive, however, and its use has declined with the increase of less-expensive plastic pipe **(Figure 4.9)**.

Plastic pipe is the least expensive type of sprinkler system piping. If plastic pipe is used, it must be listed for use in sprinkler systems **(Figure 4.10)**. It is also the lightest in weight and the easiest to install. Plastic pipe is joined with plastic cement that is listed for that purpose. The main drawback of plastic pipe is that it may not be installed in areas where the ambient temperature exceeds

Figure 4.8 Ferrous metal piping is the most commonly used type in automatic sprinkler systems.

Figure 4.9 While copper pipe is expensive, it does give an aesthetic detail to a structure when left exposed.

Figure 4.10 Plastic pipe used to supply sprinklers must be listed for that use.

specific limitations. The only plastic pipe listed for fire sprinkler systems is CPVC. CPVC is listed and approved for commercial light hazard applications as well as residential. CPVC can be installed within concealed construction or exposed as permitted within manufacturer's installation guidelines.

The wall thickness of pipe is normally referred to as the *pipe's schedule*. The **pipe's schedule** directly influences waterflow characteristics by having a larger interior diameter. Sprinkler piping is typically Schedule 10, Schedule 40, or special listed pipe.

The various types and sizes of pipes in a sprinkler system are given specific functional names based on the role each serves within the system (**Figure 4.11**). These names include:

- **Water supply main** — Piping that connects the sprinkler system to the main water supply. The water main is connected to the underground public water main that is used to supply fire hydrants and domestic use. Fire pumps can support a private water/fire main but not a public water main. The water supply main may also be connected to a private water/fire main but not a public water main. The water supply main may also be connected to a private fire main or static water source that supplies fire protection systems.

- **System riser** — Vertical piping that extends upward from the water supply to feed the cross or feed mains. The system **riser** will typically have a control valve unless an outside control valve such as a post indicator valve (PIV)

Pipe Schedule — Thickness of the wall of a pipe.

Riser — Vertical water pipe used to carry water for fire protection systems above ground, such as a standpipe riser or sprinkler riser.

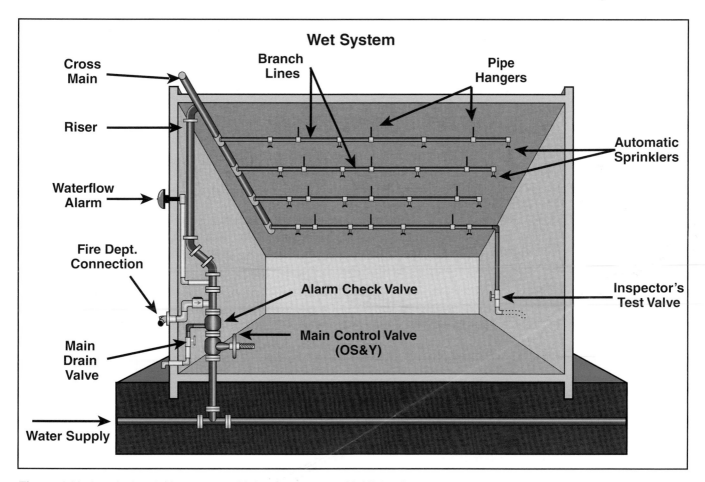

Figure 4.11 A typical sprinkler system with key components highlighted.

is used. The system riser may have an alarm check valve and will have a main drain, two pressure gauges, and a waterflow alarm device attached to it. Normally, the fire department connection is connected to the system riser above a wet system check valve and below dry, preaction, and deluge valves.

- **Sprig** — Pipe that rises vertically and supplies a single sprinkler.
- **Riser** — Any vertical supply piping in the system.
- **Feed main** — Pipe that supplies water to each of the cross mains.
- **Cross main** — Pipe that feeds the branch lines.
- **Branch lines** — Pipes that contain the individual sprinklers and include grid lines on gridded systems.

Control Valves

Every sprinkler system is equipped with various water-control valves and operating valves. Valves may be used to shut off the water, drain the system, prevent recirculation of water, and serve the system's other needs **(Figure 4.12)**.

The main control valve is used to shut off the system's water supply when it is necessary to replace sprinklers or perform other maintenance. These valves are located between the sources of water supply and the sprinkler system. After maintenance has been performed, the control valves must always be returned to the open position. Major property losses have occurred because the water supply to the sprinkler system was shut off when a fire occurred.

To help ensure that valves are not inadvertently left in a closed position, all valves that control water to sprinkler systems are of the indicating type. These valves must be supervised by manual or electronic means. For example, valves must be chained and padlocked in the open position and/or electronically connected to the fire alarm. With an indicating valve, the position of the valve — open or closed — can be determined at a glance. All aboveground control valves installed that are capable of isolating water supply to sprinklers must be indicating. Underground valves in water distribution systems are commonly nonindicating. The three types of control valves are the gate, butterfly, and ball valves.

NOTE: Floor control pressure reducing valves are required to be indicating, but the indicated position only applies to the hand wheel. It is not an indication of the valve's position relative to the seat. A valve indicating "OPEN" is actually shut unless water is flowing downstream.

Gate Valves

The three basic types of gate valves are the outside stem and yoke (OS&Y), the post indicator valve (PIV), and the wall post indicator valve (WPIV).

- OS&Y valves have a yoke on the outside with a threaded stem that controls the opening and closing of the gate by turning a hand wheel **(Figure 4.13)**. The threaded portion of the stem is outside the yoke when the valve is open and inside the yoke when the valve is closed. The position of an OS&Y valve is easily seen from a distance as opposed to other types of indicating valves.
- The PIV is used to control underground sprinkler valves and consists of a hollow metal post attached to the valve housing **(Figure 4.14)**. The valve

Figure 4.12 Numerous valves are incorporated into a sprinkler system to serve specific purposes.

Figure 4.13 An outside stem and yoke (OS&Y) valve shows the threaded stem when the valve is open.

Figure 4.14 A typical PIV.

stem is inside this post. Mounted on the stem is a movable target with the words "OPEN" or "SHUT" visible through a window depending on the position of the valve. The operating handle is fastened and normally locked to the post. PIVs may be of the indicating butterfly type. These valves have a paddle indicator or a pointer arrow that shows the position of the valve.

- The WPIV is similar to the PIV, except that it extends through the building wall with the target and valve-operating wheel on the outside of the building (**Figure 4.15, p. 118**). A gate-valve mechanism consists of a close-tolerance gate that slides across a waterway (**Figure 4.16, p. 118**).

Figure 4.15 Multiple WPIVs in use at a big-box store.

Figure 4.16 To shut the valve, the gate valve slides down to close the waterway.

Butterfly Valves

In a butterfly valve, a disc rotates 90 degrees inside the waterway. Butterfly valves are operated by a worm gear, which is turned by a handle or a handwheel **(Figure 4.17)**. The position of a butterfly valve is indicated by a pointer that points either to the "OPEN" or the "CLOSED" setting on the valve body or by a cross-view of the valve showing its open and closed position.

Ball Valves

A ball valve may be used as a control valve as long as it cannot be closed in less than five seconds **(Figure 4.18)**. Closing any valve in less than five seconds is ill-advised because it may cause water hammer, resulting in damage to the system components. Ball valves that are listed for use in a fire sprinkler system as a control valve have a handwheel. These valves are typically found as floor-control valves. The ball valve has an indicating device on the body of the valve indicating whether the valve is open.

Auxiliary Valves

Auxiliary valves can be either manually operated or automatic and are provided as relief, drain, test, or air service valves. Manually operated auxiliary valves are permitted to be nonindicating valves when installed in piping as not to isolate the water supply from sprinklers. Automatic auxiliary valves relieve excess pressure, drain water, or vent air from the system. Examples of auxiliary valves follow.

Figure 4.17 The butterfly valve rotates inside the waterway.

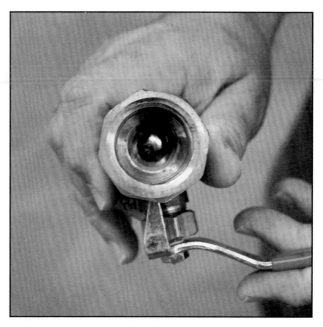

Figure 4.18 To prevent water hammer, ball valves must not be closed in less than five seconds.

Automatic Drain Valves

Automatic drain valves (also known as *ball drip valves*) drain piping when pressure is relieved in the pipe. The most common application of these valves is to drain water from Siamese connections or the FDC after use in certain types of sprinkler systems and in the trim piping of dry pipe systems. This technique prevents the portion of the pipe that extends through the wall from freezing in cold weather.

Globe Valves

Globe valves are small handwheel-type valves that are primarily used on drains and test valves **(Figure 4.19)**. These valves include the inspector's test valve, which is used to cause water to flow through the system for an inspection test. The inspector's test valve may be located near a riser, for convenience, or at a remote location within the sprinkler system.

Stop or Cock Valves

These valves are also used for drains and alarm testing. Ball-type valves are most commonly opened and closed with a quarter turn of the valve.

Relief Valves

Relief valves must automatically relieve excess water pressure or vent air from systems when required by NFPA® 13. The relief valve is provided to open and discharge water to drain if the pressure becomes excessive.

Figure 4.19 Globe valves are primarily used on drains and test valves.

Check Valves

Check valves are used to limit the flow of water in one direction (**Figure 4.20**). They are placed in water sources to prevent recirculation or backflow of water from the sprinkler system into the municipal water supply. These valves are also required in the FDC line to prevent water from flowing out of the sprinkler system through the FDC. Check valves usually have an arrow cast into the body to indicate the direction of flow. If there is no arrow or other indicator, the valve clapper pivot should be on the end toward the source of supply.

Figure 4.20 A check valve ensures the proper direction of waterflow during operations that use a fire department connection.

Sprinklers

A heat-activated sprinkler is the portion of the sprinkler system that reacts to the heat of a fire and then discharges water onto the fire area. In addition to heat-activated sprinklers, there are open sprinklers in deluge-type systems that have no fusible element.

A simple thermally-sensitive device controls most automatic sprinklers. The fusible link and frangible bulb are two common types of these devices (**Figures 4.21a and b**). In its simplest form, the fusible link is a solder link

Figure 4.21a A fusible link sprinkler.

Figure 4.21b A frangible bulb sprinkler.

with a low, precisely established melting point. The solder link is connected to a cap that restrains the water at the orifice. When there is a fire, the solder in the fusible link melts at its predetermined melting point. The lever arm or struts are released and spring clear of the sprinkler frame **(Figure 4.22)**. As the lever arm drops, the seated cap is released, which permits the water to flow.

Frangible bulbs are inserted between the sprinkler frame and the discharge orifice in much the same way as the fusible link. Frangible bulbs contain a liquid that expands when heated. When the expansion of the liquid within the bulb exceeds the bulb's strength, the bulb breaks, allowing water to flow.

Sprinklers can be modified for different environments. For example, if a sprinkler is to be placed in a corrosive atmosphere, such as a plating room, it can be coated with wax for protection and listed for such use by the manufacturer. If mechanical damage is possible, the sprinkler can be fitted with a protective cage **(Figure 4.23)**. In areas where appearance is important, such as in a residential setting, a variety of recessed, flush, and concealed sprinklers are available that have a finish matching the color and texture of the ceiling **(Figure 4.24)**. The sprinkler manufacturer, not the contractor or occupant, must apply such finishes.

Figure 4.22 As the solder in the fusible link melts, the lever arms or struts are released.

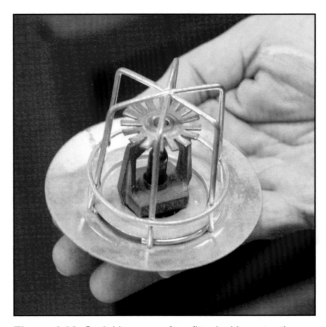

Figure 4.23 Sprinklers are often fitted with protective cages in areas where they may become damaged.

Figure 4.24 A typical recessed sprinkler.

Figure 4.25 Traditional FDCs have two 2½-inch (65 mm) female inlets to which fire hoses can be connected.

Figure 4.26 Some FDCs incorporate inlets for large diameter hose (LDH).

Sprinklers have certain characteristics, such as the various temperature ratings, response times, orifice diameters, and deflector types. These characteristics will be explained later in this chapter.

Fire Department Connection (FDC)

Commercial and certain residential sprinkler systems are required to have one or more fire department connections (FDC). These connections permit a fire department pumper to pump into the sprinkler system and supplement the water supply, boosting the pressure within and the volume of water discharged from the system. Fire department connections are especially important to the protection of a building in which:

- The primary water supply pressure is weak.
- The supply is being overtaxed by a large number of open sprinklers.
- The system is compromised by obstructed/occluded piping or partially closed control valves.

The most common type of FDC connection is a 4-inch (100 mm) pipe equipped with a fitting on the outside of the building. The fitting has two 2½-inch (65 mm) female inlets to which fire hoses can be connected (**Figure 4.25**). Larger systems may have three or more inlets. Some newer connections may be equipped with inlets for large diameter hose (LDH) (**Figure 4.26**). Fire department connections and their requirements are discussed in more depth in Chapter 5, Standpipe and Hose Systems.

Fire Department Connections

FDCs supporting sprinkler systems are required only to supplement sprinkler demand and are not sized to provide the total required flow. FDCs serving standpipes are required to be sized to support total standpipe design demand.

Pipe Supports

Pipe supports consist of three elements: a component that attaches to the structure, a component that attaches to the sprinkler pipe, and a component that links the two together. As the name implies, pipe supports are intended to support the sprinkler piping. A hanger is the most common pipe support.

Hangers

Hangers are a listed and approved method or engineered certified system of attaching sprinkler piping to the structure. Hanger size and spacing are primarily determined by the diameter of the pipe it is supporting, but it is also affected by segment lengths and pipe material. In all cases, hangers supporting sprinkler piping must be capable of supporting five times the weight of the water-filled pipe, plus 250 pounds (125 kg). Where the water pressure exceeds 100 psi (700 kPa), NFPA® 13 requires additional hangers and vertical-lift restraints. Unless engineered otherwise, components supporting sprinkler systems shall only support the sprinkler pipe and sprinkler components. No other building system, wiring, or equipment is permitted to be supported or attached to the sprinkler system.

Seismic Restraints and Bracing

Sprinkler systems must be protected from unusual seismic or blast events. Similar to hanger pipe supports, seismic protection requires components to be listed and approved or engineer certified when provided to attach sprinkler piping to the structure. However, seismic protection does not support the weight of the piping as pipe hangers do. Seismic protection uses a combination of flexible pipe couplings, increased clearances around pipe penetrations through walls and floors, and longitudinal and lateral sway bracing and restraints to protect sprinkler pipe against lateral and horizontal movement.

Important Aspects of Sprinklers

Sprinklers, sometimes referred to as *sprinkler heads* or simply *heads*, are the part of the suppression system that discharges water onto the fire area. The main characteristics of a sprinkler are:

- Temperature ratings
- Response times
- Orifice diameter
- Deflector components

Listings and Approvals

The selection of the operating temperature of a sprinkler is determined based upon the maximum air temperature expected at the level of the sprinkler under normal conditions. Typical operating temperatures vary from 135°F to 650°F (57°C to 343°C). For ordinary room temperatures, a sprinkler with a temperature rating of 135°F to 170°F (57°C to 77°C) is most frequently used. If the typical ambient temperature exceeds 100°F (38°C), such as in an attic or near a heater, a sprinkler with a higher temperature rating is used. It is necessary to provide a margin between normal room temperature and the operating temperature in order to prevent inadvertent activation.

Temperature ratings that are too low for a given location may result in malfunction or accidental activation. Temperature ratings that are too high will delay sprinkler operation, enhancing fire growth. However, it is common to see high temperature (blue) sprinklers in warehouses. Testing and experience have revealed that the use of high temperature sprinklers can reduce the number of sprinklers that open and the volume of water needed to achieve fire control in many high-challenge occupancies. Sprinkler frames or glass bulbs are color-coded so that their temperature ratings can be quickly distinguished **(Figure 4.27)**.

For example, those sprinklers rated for temperatures from 250°F to 300°F (121°C to 149°C) are blue. The temperature rating is also stamped on the link in fusible

Figure 4.27 Sprinkler glass bulbs or frames are color-coded so that their temperature ratings can be quickly distinguished.

link-type sprinklers. On other types of sprinklers, the temperature rating is stamped on some other part of the sprinkler.

Sprinkler Response Time

Because most sprinklers are thermal devices, there is some delay between the ignition of a fire and the operation of the sprinkler. This delay is a function of several variables, including the sprinkler's design.

The activation time of the sprinkler depends on the surface area, mass, and thermal characteristics of the heat-sensitive element. Although the gas temperature surrounding a sprinkler may be 165°F (74°C), a heat-sensitive element rated at 165°F (74°C) will not reach this temperature for some time. In addition, the temperature and velocity of a gas jet as it travels across the ceiling will affect the activation time because heat must be transferred from the hot gas jet to the operating elements of the sprinkler. The faster and hotter the gas jet, the faster the sprinkler will operate.

To speed the operation of sprinklers, engineers have designed types of sprinklers known as fast-response or quick-response sprinklers. These faster-operating sprinklers can be compared to ordinary sprinklers through the use of a Response Time Index (RTI). These sprinklers typically react 5 to 10 times faster than traditional sprinklers. The lower a sprinkler's RTI, the faster it responds. Examining the operating element can typically identify fast-response and quick-response sprinklers. If glass bulbs are used, the bulbs will be much thinner than standard response bulbs. If metallic links are used, the links will be of almost foil-thin metal.

Orifice Diameter

Sprinklers are also manufactured with different discharge orifice diameters that play a part in a sprinkler's **K-Factor** rating. For any given pressure, the amount of water that can be delivered from a sprinkler is dependent upon the orifice size. Sprinkler orifice diameters can range from ¼ inch (6.4 mm) to 1 inch (25 mm). The most common diameter is ½ inch (12.7 mm).

Deflector Component

Another important component of a sprinkler is the deflector, which is attached to the sprinkler frame and creates the discharge pattern of the water **(Figure 4.28)**. Pressure forces water against the deflector to convert it into a spray pattern.

Sprinklers produced before 1955 (commonly known as *old-style sprinklers*) were designed with deflectors that discharged a large portion of the water upward (about 40 percent) toward the ceiling in order to protect structural elements. Unfortunately, this design did not produce a good downward distribution of water. Modern standard sprinklers produce a more uniform discharge pattern that is directed downward. By directing all of the water downward, the fire is controlled more effectively, resulting in reduced ceiling temperatures and better protection for structural elements. Because of the difference in discharge patterns, old-style sprinklers cannot be used to replace modern sprinklers. Modern sprinklers may be substituted for old-style sprinklers in an existing system if it becomes necessary to change or upgrade a sprinkler system.

K-Factor — Discharge coefficient used in hydraulic calculations of sprinkler systems and is based upon the sprinkler discharge orifice and physical characteristics of the sprinkler.

Figure 4.28 The deflector is attached to the sprinkler frame and creates the discharge pattern of the water.

The deflector configuration is fundamental to the effectiveness of the sprinkler. Basic sprinkler deflector types include the following:

- **Upright** — Designed to deflect the spray of water downward in a hemispherical pattern **(Figure 4.29)** Upright sprinklers cannot be inverted for use in the hanging or pendent position because the spray would be deflected toward the ceiling.

- **Pendant** — Used where it is impractical or unsightly to use sprinklers in an upright position such as below a suspended ceiling **(Figure 4.30)**. The deflector on this type of sprinkler breaks the pattern of water into a circular pattern of small water droplets and directs the water downward.

Figure 4.29 An upright sprinkler.

Figure 4.30 Pendant sprinklers are used in applications where an upright sprinkler is not practical.

- **Sidewall** — Used in instances where it may be desirable or required to install sprinklers on the wall at the side of a room or space **(Figure 4.31)**. This approach may be for cost-savings or appearance. By modifying the deflector, a sprinkler can be made to discharge most of its water to one side. Sidewall sprinklers are useful in areas such as corridors, offices, hotel rooms, and residential occupancies.

- **Concealed** — Hidden by a removable decorative cover that releases when exposed to a specific level of heat **(Figure 4.32)**.

- **Flush** — Mounted in a ceiling with the body of the sprinkler, including the threaded shank, above the plane of the ceiling.

- **Recessed or semi-recessed** — Installed in recessed housing within the ceiling of a compartment or space; all or part of the sprinkler other than the threaded shank is mounted in the housing.

Figure 4.31 Sidewall sprinklers are commonly found in corridors and hotel rooms.

Figure 4.32 A hidden/concealed sprinkler.

Specialty Sprinklers and Nozzles

While the types of sprinklers previously listed can be used in a variety of circumstances, there are situations where a special type of sprinkler may be best suited or required by the codes adopted by the authority having jurisdiction (AHJ). These might be situations that call for a particular pattern of water distribution or unusual ambient temperatures that mandate special construction. Common types of specialty sprinklers include:

- **Control Mode Specific Application (CMSA)** — Produces large drops of water to enable the spray to penetrate strong updrafts created by elevated fire conditions.

- **Dry Sprinkler** — Used in either dry or wet systems where the protected area may freeze. An example would be a walk-in freezer in a store or storage facility. These sprinklers can be installed in the upright, pendent, or sidewall position.

- **Residential** — Fast-response sprinklers that have a special low-mass fusible link or bulb that makes the time of temperature actuation much less than that of a conventional sprinkler. These sprinklers are listed for use in residential occupancies.

- **Attic** — Designed to protect sloped attic spaces. The attic discharge patterns were developed to produce a narrow but long pattern **(Figure 4.33)**. One of its primary functions is to provide protection of the roof structure.

Figure 4.33 Attic sprinklers are designed to provide protection to roof structures.

- **Water Mist** — Turns the water into a fine mist, which depletes oxygen and blocks radiant heat. There are several different water-mist sprinkler designs available that can be matched for the specific hazard or application.

- **In-Rack** — Typically used in storage facilities. In-rack sprinklers incorporate a protective disc that shields the heat-sensing element from water that is discharged from the sprinklers above **(Figure 4.34)**.

- **Early Suppression Fast Response (ESFR)** — Engineered to combat the growing challenge associated with high-rack storage facilities. These sprinklers combine the advantages of fast response with effective and efficient application of large quantities of water in high challenge fire scenarios.

Figure 4.34 An in-rack sprinkler.

- **Extended Coverage Sprinklers (EC)** — Similar to standard spray sprinklers in all respects, except for its listed coverage areas. These sprinklers are designed to throw water further and cover a greater area per sprinkler while still supplying the water distribution necessary for suppression.

- **Open Sprinklers** — Used with deluge systems and resemble standard sprinklers without the thermal element, which keeps them closed. These sprinklers are always open. Every sprinkler within a deluge system will discharge water if the system is activated rather than just the sprinklers exposed to heat.

Types of Sprinkler Systems

While the components of all sprinkler systems are basically the same, there are different types of sprinkler systems depending upon the needs and requirements for the occupancy and environment. The most common types of sprinkler systems in use today include the following:

- Wet-pipe systems
- Dry-pipe systems
- Deluge systems
- Preaction systems

Wet-Pipe Systems

A wet-pipe sprinkler system contains water under pressure in the piping at all times so that the opening of a sprinkler immediately discharges water onto a fire and activates an alarm **(Figure 4.35)**. Wet-pipe sprinkler systems are popular due to their simplicity and reliability and are usually successful in controlling fires with only a minimum number of sprinklers opening. NFPA® research indicates that 99 percent of fires are controlled with less than 10 sprinklers activated.

Depending on the occupancies in which they are built, wet-pipe systems fall under the requirements of NFPA® 13, 13D, or 13R. Standard wet-pipe systems may be installed in any location where system components will not be subject to freezing temperatures unless protected from freezing in accordance with

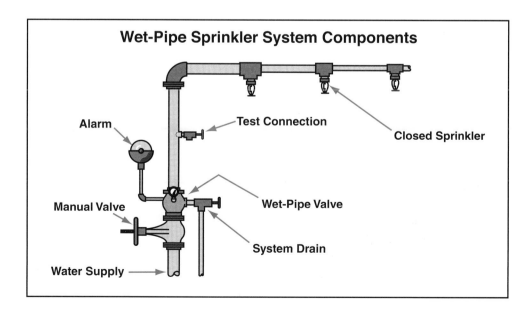

Figure 4.35 Wet-pipe sprinkler systems contain water under pressure in the piping at all times.

NFPA® 13. There are two types of wet-pipe systems: one type uses an alarm-check valve and the other type uses a waterflow-detecting alarm device to indicate waterflow. Either system can incorporate antifreeze where the piping is subject to freezing.

Alarm Check Valves

All alarm check valves have the following in common: trim, clapper, and drain valve. Alarm and test valves are installed on the trim, and the main drain valve is located above the clapper. When water flows downstream of the riser or on pressure surges, the clapper rises off its seat, allowing water to flow into the retard chamber and activate an alarm when waterflow above the clapper is persistent.

Waterflow Switch Device

The other variety of wet systems is typically equipped with a waterflow-detecting alarm device that initiates an alarm when water begins to flow in the system **(Figure 4.36)**. A waterflow device consists of a vane or paddle that protrudes through the riser into the waterway. The vane is connected to an alarm switch on the outside of the pipe. Movement of the vane, caused by water flowing at a rate of at least 10 gpm (40 L/min), operates the switch that initiates an alarm. The vane must be thin and pliable so that if many sprinklers operate, the waterflow will flatten the vane against the wall of the riser, resulting in a clear flow of water through the pipe.

Just as the **retard chamber** helps prevent false alarms with an alarm check valve, there is a time-delay feature on waterflow switches. This delay is typically 30 seconds but may be up to 90 seconds from activation depending on the frequency and severity of water pressure surges and AHJ requirements.

Antifreeze Systems

Where buildings or portions of buildings are subject to freezing temperatures, wet systems may be used if the water is replaced by an antifreeze solution. These systems are called antifreeze systems. In these systems, once a sprinkler operates, the antifreeze solution is discharged and water enters the system. Increased concern for environmental health has resulted in regulations restricting the type of chemical that can be used in any piping system connected to a public water supply. However, certain grades of glycerin or propylene glycol are usually acceptable. In systems not connected to a public water supply, other liquids can be used. Some antifreeze solutions increase the fire hazard, and current standards only permit factory premixed solutions.

Retard Chamber — A collection chamber used on wet systems with alarm check valves whose function is to prevent false alarms due to water pressure surges.

WARNING!
Antifreeze is combustible. Use only listed solutions premixed at the factory.

Figure 4.36 Waterflow indicators initiate an alarm when water movement is detected in the system.

Figure 4.37 A U-loop piping arrangement prevents the antifreeze solution from leaking back into the rest of the system.

Extensions may be added to a wet system and filled with an antifreeze solution to protect small areas, such as loading docks that are subject to freezing. These extensions are often referred to as "antifreeze loops" and are constructed in the shape of a U-loop **(Figures 4.37)**. This arrangement prevents dilution of the antifreeze solution and keeps the antifreeze out of the main portion of the system.

Reduced pressure zone backflow preventers may be required to ensure nothing from an antifreeze system reaches the potable water supply. In addition, a listed expansion tank may be required.

Dry-Pipe Systems

A dry-pipe sprinkler system is one in which air or nitrogen under pressure replaces water in the pipes. When buildings lack sufficient heat to keep the water in the sprinkler pipes from freezing, a dry-pipe sprinkler system must be used. The dry-pipe valve must be located in a heated portion of the structure, which must be maintained at a minimum of 40°F (4°C) according to code requirements. If the valve is subject to freezing conditions, it may not activate properly when required. In some cases, it may be necessary to build a small, heated enclosure specifically for the dry-pipe valve riser and its associated equipment.

A dry-pipe valve uses the force created by the air pressure in the system to hold the clapper in the valve closed and keep water from entering the system until a sprinkler operates **(Figure 4.38)**. When a sprinkler opens, there is a reduction in the system air pressure and the dry-pipe valve opens. This reduction admits water into the sprinkler system.

For a dry-pipe system to operate, heat from a fire must cause the sprinkler to activate. The pressurized air or nitrogen contained in the piping will flow out through the open sprinkler. Once the pressure in the piping is significantly reduced, the dry-pipe valve trips open. Water will then flow through the piping system and be discharged through the open sprinkler. In large dry systems, quick opening devices may be provided to speed up the operation of the dry valve.

The air or gas necessary to service a dry system may be obtained from several sources. One source is an air compressor arrangement dedicated for exclusive use with the system. A second alternative is to have the sprinkler system connected to a plant or shop air system from a reliable source. Though not as common, compressed air or nitrogen gas cylinders may also be used to provide the pressure within the dry system. Nitrogen gas has the added benefit of reducing the rate of piping corrosion.

The dry-pipe valve uses a differential principle incorporated into the valve design to open. When the differential dry-pipe valve operates and the clapper swings open, a latch holds the clapper in place to prevent it from reseating. Releasing the latch in some valve models requires that the valve body be opened. Newer valves with an external reset feature allow resetting the clapper without having to remove the faceplate. Therefore, these systems must be manually reset after they have activated.

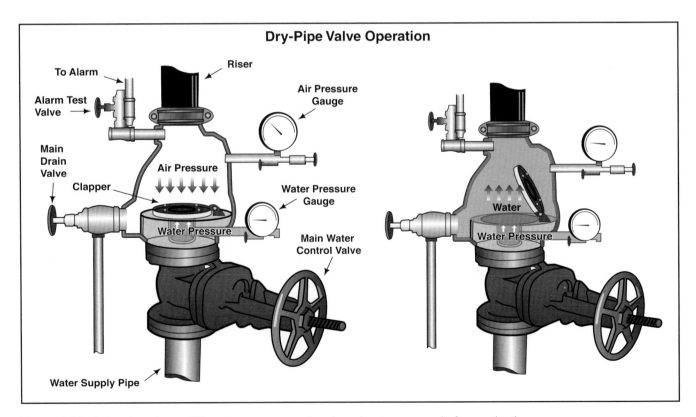

Figure 4.38 A dry pipe shows different pressures on the air and water gauges before activation.

The air-water differential allows a small amount of air pressure to hold back the water. This differential is a ratio of air pressure to water supply pressure within the system to maintain the dry-pipe valve in the closed system. The differential is generally between a 3:1 to 7:1 ratio, where, for example, 1 psi (7 kPa) of air is needed to hold back 3 psi (20 kPa) of water. In addition, the less air pressure in the system means the quicker water can expel the air and discharge water on the fire. The air pressure in the system must be maintained within the prescribed limits, which are typically no more than 20 psi (140 kPa) above the pressure at which the dry-pipe valve will trip.

The dry-pipe valve has unique design characteristics when compared to wet systems. First, it does not open immediately upon activation of a sprinkler. Second, when the air pressure is depleted, the clapper releases, allowing water to enter the system.

Dry-pipe systems can use quick-opening devices to help speed up the process of getting water to the source of the fire. In large dry-pipe systems, several minutes can elapse as air is being expelled through the open sprinklers before water is discharged. Quick-opening devices accelerate the trip time, reducing the delay of the water application. Accelerators and exhausters are two types of quick-opening devices. An accelerator works by unbalancing the differential in the dry-pipe valve, which causes it to trip more quickly. The accelerator detects a drop in system pressure and forces the clapper open to allow water into the system. The second variety is the exhauster, which is no longer approved and listed but may be found on older systems. An exhauster functions by quickly expelling air from the system. The exhauster is generally located at the most remote end of the system from the water supply and does not work as fast as the accelerator.

Quick-opening devices are required on any dry-pipe system with a capacity of more than 500 gallons (1 900 L) if the system cannot meet the water delivery time requirement (water must be discharged from the system test connection in not more than a specified duration after opening the inspector's test connection, starting with the system at normal air pressure), and on any system with a capacity greater than 750 gallons (2 800 L). Where approved by the AHJ, water delivery time calculations may be used in lieu of the trip-time requirements. When using the calculation method, water delivery times must meet the hazard class requirements.

Deluge Systems

A deluge system is designed to quickly supply a large volume of water to the protected area. Deluge systems are normally used to protect extra-hazard occupancies such as:

- Aircraft hangars
- Woodworking shops
- Cooling towers
- Ammunition storage
- Certain manufacturing facilities

Deluge systems are sometimes used with foam to protect against flammable liquid hazards **(Figure 4.39, p. 132)**. Because these systems require large volumes of water, fire pumps sometimes supply them.

Figure 4.39 A system of pipes, sprinklers, and control valves are designed to automatically discharge water or extinguishing agent to prevent the spread of fire while first responders travel to the scene. *Courtesy of the U.S. Navy.*

A deluge system is similar to a dry-pipe system in that no water is contained in the piping before the activation of the deluge valve. However, the deluge system differs from a dry-pipe system in that all of the sprinklers are open with no fusible links. This difference means that when the valve is tripped and water enters the system, the water will simultaneously discharge through all the sprinklers. A deluge valve controls the flow of water to the system. Fire detection devices, which may be heat, smoke, and/or flame detectors, are installed in the same area as the sprinklers. The detection devices control the operation of the deluge valve through a tripping device and are required to be automatically supervised. Unlike wet- or dry-pipe systems, the sprinklers do not function as the detection device in a deluge system.

Just as there are several types of fire protection systems, several methods of operating the deluge valve exist. They can be operated electrically, pneumatically, or hydraulically. Deluge valves also have provisions for manual operation. Some operations are described as follows:

- **Electrical operation** — Deluge valve designed for use with fire detectors that transmit an electrical signal to the valve. The activation system includes a releasing solenoid that opens the deluge valve clapper. A manual release and reset in case of power failure is included. Normal electrical service is a primary power source; however, batteries are usually provided as a secondary power source. A major advantage of the electrically operated deluge valve is its speed of operation. Most modern deluge systems are activated electronically and use similar components to a fire alarm system.

- **Pneumatic operation** — Deluge valves designed for use in a pneumatic detection system, usually one that has rate-of-rise detectors. This type of activation is actually a combination of pneumatic and mechanical activation. A pneumatic detector operates on an imbalance of pressure. The deluge valve activates when a change in air pressure within the device (caused by an increase in temperature) displaces a diaphragm. The diaphragm is mechanically linked to a tripping mechanism that, in turn, releases the deluge valve clapper. The pneumatic system does not require electrical power for activation, but a manual release is required. Another variety of pneumatic system uses a "pilot line" which is a separate and independent system of small diameter piping supplying closed sprinklers located in the area to be protected by the deluge system. This piping is filled with compressed air and the opening of a sprinkler on the pilot line results in a loss of pressure, which will cause the deluge valve to open.

- **Hydraulic operation** — There are several different types of hydraulically operated deluge valves. Some of these valves may use a combination of hydraulically operated valves and a pneumatic, hydraulic, or electrical detection system. Regardless of the detection system, a release in hydraulic pressure is usually required for activation of the deluge valve.

In a situation where an individual discovers the fire, all deluge valves can be manually activated. As water enters the deluge valve and the piping, a pressure switch (a water motor gong) is activated that transmits an alarm either locally or to a supervising station. The water then flows simultaneously through all open sprinklers.

Preaction Systems

A preaction system is used when it is important that the inadvertent release of water be minimized, even if the sprinkler pipes accidentally break. Preaction systems are frequently used to protect computer rooms, document-storage areas, freezers, and cold-storage warehouses. The system employs a deluge valve, identical to the types used in deluge systems. These valves will have the word "Deluge" stamped on the valve housing. These systems also will have a detection system, but all of the sprinklers will be closed. Like deluge systems, the preaction system valve will not discharge water into the system's piping until an indication is received from fire detection devices that a fire may exist. Once water is in the system, it may then be discharged through any sprinkler that has opened.

The piping in a preaction system is dry, and it can be used in freezing environments such as cold-storage warehouses. Where the preaction system exceeds 20 sprinklers, regulations require that the piping be supervised. This requirement is usually accomplished with air under low pressure. In the event of a leak or break in the piping, the air pressure drops and transmits an alarm without admitting water into the system. Three basic categories of preaction sprinkler systems are as follows:

- **Noninterlock system** — Admits water to sprinkler piping upon operation of detection devices or automatic sprinklers.

- **Single interlock system** — Admits water to the sprinkler piping upon operation of a detection device.

- **Double interlock system** — Admits water to sprinkler piping upon operation of both a detection device and a loss in air pressure caused by automatic sprinkler activation.

Activation of a fire alarm device, such as a pull station or a smoke detector, sends a signal to the preaction valve that causes the valve to open. Pressure switches at the deluge valve detect the flow of water into the system and trigger the waterflow fire alarm. When the level of heat at a sprinkler reaches the appropriate temperature, the sprinkler opens and water flows through the orifice.

Water-Mist Fire Suppression Systems

The water-mist system is gaining more popularity in the field of fire protection. While not considered an automatic sprinkler system, a water-mist fire suppression system is similar. This system sprays water onto a fire in a fine mist that absorbs larger quantities of heat than an automatic sprinkler system. The fine mist of water controls or extinguishes the fire by cooling and blocking radiant heat production. A benefit of water-mist fire suppression systems is the limited amount of water used compared to traditional sprinkler systems. These systems prevent and mitigate excessive water damage in occupancies where they are used.

The development of water-mist systems is relatively recent and has been driven by the need for a replacement for halon and for other locations where traditional sprinkler applications are not conducive with the hazard. See Chapter 7 for additional information on halon. In addition, there has been a need to provide sprinkler protection on passenger cruise ships. NFPA® 750, *Standard on Water Mist Fire Protection Systems*, provides the requirements for installation of water-mist fire suppression systems. Water-mist systems may be found in locations such as tunnels, passenger ships, aircraft passenger compartments, machinery spaces, and turbine enclosures. These applications are limited to the size of the area protected and can be applied to local application, total compartmentation, or zoned application for Class A, Class B, and Class C fires.

In general, a water-mist system is composed of small-diameter, pressure-rated copper, or stainless-steel tubing. Small-diameter spray nozzles are spaced evenly on the tubing. Depending on the design of the system, the spray nozzles may be of the open or closed sprinkler variety. Systems are activated by a product-of-combustion detection system. Water-mist systems should also have a means of manual operation should individuals discover the fire before the activation of the system.

Water-mist systems are designed to be operated at considerably higher pressures than standard sprinkler systems. These systems are designed for fire suppression, control, or extinguishment purposes and may have a set amount of time of discharge. Water-mist systems are divided into low [175 psi (1 200 kPa) or less], intermediate [175 to 500 psi (1 200 to 3 500 kPa)], or high [500 psi (3 500 kPa) or greater] pressure systems. Compressed-air, nitrogen, or high-pressure water pumps create these higher pressures. The most common type of water-mist system works similarly to a traditional deluge sprinkler system. All of the spray nozzles in a particular room or zone are open, and when the detection devices activate, the water is allowed to discharge from the spray nozzles.

Water-mist systems may present an eye hazard during operation due to the high operating pressure of the system. These high-pressure systems may also present a noise hazard during discharge. Additionally, the activation of water-mist systems can result in reduced visibility, which may increase the time necessary for egress of occupants from the protected area.

Foam Fire Suppression Systems

A foam fire suppression system is used when water alone may not be an effective fire suppression agent. In general, foam works by forming a blanket on the burning fuel. The foam blanket excludes oxygen, stops the burning process, and cools adjoining hot surfaces. The type of system and foam used depends largely on the hazard being protected. These hazards and locations include but are not limited to:

- Flammable liquids
- Processing facilities
- Storage facilities
- Aircraft hangars **(Figure 4.40)**
- Special hazard facilities

Foam extinguishes a fire by one or more methods, including:

- **Smothering** — Prevents air and flammable vapors from combining.
- **Separating** — Intervenes between the fuel and the fire.
- **Cooling** — Lowers the temperature of the fuel and adjacent surfaces.
- **Suppressing** — Prevents the release of flammable vapors.

Figure 4.40 Areas such as aircraft hangars are often protected by a high expansion foam system.

Standards and requirements for foam system design, placement, and other technical information can be found in the following:

- NFPA® 11, *Standard for Low-, Medium-, and High-Expansion Foam*
- NFPA® 16, *Standard for the Installation of Foam-Water Sprinkler and Foam-Water Spray Systems*
- NFPA® 30, *Flammable and Combustible Liquids Code*
- NFPA® 409, *Standard on Aircraft Hangars*

Design specifications and requirements of foam sprinkler systems can also be found in publications from manufacturers and nationally recognized testing laboratories, such as Underwriters Laboratories Inc. (UL) and FM Global.

Types of Foam

There are a number of types of foam agents available. Some foam concentrates are designed for specific applications, while others are suitable for all types of flammable liquids. The most common types of foam include the following:

- Aqueous film forming foam (AFFF)
- Fluoroprotein (FP)
- Film-forming fluoroprotein (FFFP)
- Protein (P)

- Alcohol-resistant (AR)
- Medium- and high-expansion foam
- Class A Foam

Foam Generation

Most fire-extinguishing foam concentrates in use today are of the mechanical type, which means they must be proportioned with water and mixed with air (aerated) before they can be used. Before discussing the foam-generating process, it is important to understand the following terms:

- **Foam concentrate** — Raw foam liquid before the introduction of water and air. It is usually shipped in 5-gallon (19 L) buckets or 55-gallon (200 L) drums but may be shipped in totes that contain several hundred gallons (liters) of product. Concentrate can also be stored in large fixed tanks that can hold 500 gallons (1 900 L).
- **Foam proportioner** — Device that introduces the correct amount of foam concentrate into the water stream to make the foam solution.
- **Foam solution** — Homogeneous mixture of foam concentrate and water before the introduction of air.
- **Foam (also known as *finished foam*)** — Completed product once air is introduced into the foam solution.

Four elements are necessary to produce high-quality fire fighting foam:

- Foam concentrate
- Water
- Air
- Mechanical agitation

All of these elements must be present and introduced in the correct ratios. Removing any element will result in either no foam or an unusable liquid.

The formation of foam involves two stages:

- Water is mixed with the foam liquid concentrate to form a foam solution, which is known as the *proportioning stage* of foam production.
- The foam solution passes through the pipeline or hoseline to a foam nozzle, sprinkler, or generator, where the solution is aerated to form finished foam product.

An eductor is a type of foam proportioner that injects foam concentrate directly into the water flowing through a hose or pipe **(Figure 4.41)**. The eductor is designed to inject the correct amount of foam concentrate so that the desired consistency of foam solution is achieved. There are many different types of eductors, and some of these types will be explained later in the chapter.

Proportioning equipment and foam nozzles, sprinklers, or generators are engineered to work together. Using a foam proportioner that is not hydraulically matched to the foam nozzle, sprinkler, or generators, even if the two are made by the same manufacturer, can result in unsatisfactory foam or no foam at all.

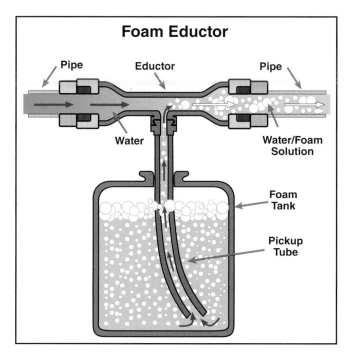

Foam Eductor

Pipe Eductor Pipe

Water

Water/Foam Solution

Foam Tank

Pickup Tube

Figure 4.41 As water flows past the pickup tube orifice, foam concentrate is drawn into the water stream to create the foam solution.

Types of Foam Systems

A foam system must have an adequate water supply, a supply of foam concentrate, a piping system, proportioning equipment, and foam makers (or discharging devices). Each system is discussed in further detail in the sections that follow:

- Fixed-Foam
- Semifixed Type A
- Semifixed Type B
- High-Expansion Foam
- Foam/Water

Fixed Foam

A fixed-foam fire-extinguishing system is a complete installation that is piped from a central foam station **(Figure 4.42, p. 138)**. These systems automatically discharge foam through fixed delivery outlets to the protected hazard. If a pump is required to increase pressure in the system, it is usually permanently installed. Fixed systems may be the total-flooding or local-application type. Most fixed systems are deluge types that have unlimited water supplies and may have large foam supplies. They require actuation by some sort of product-of-combustion detection system. Fixed systems use low-, medium-, or high-expansion foam.

Semifixed Type A

In a semifixed Type A system, the foam discharge piping is in place but is not attached to a permanent source of foam. The semifixed Type A system requires a separate mobile foam-solution source, which is usually a fire brigade or fire department pumper **(Figure 4.43, p. 138)**. This type of system is found in settings that involve several similar hazards, such as petroleum refineries, and are used primarily on flammable liquid storage tanks.

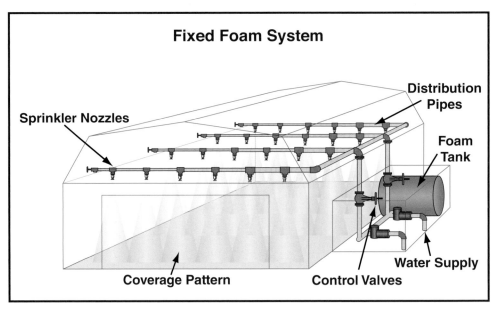

Figure 4.42 In a fixed foam system, foam is discharged directly from fixed outlets.

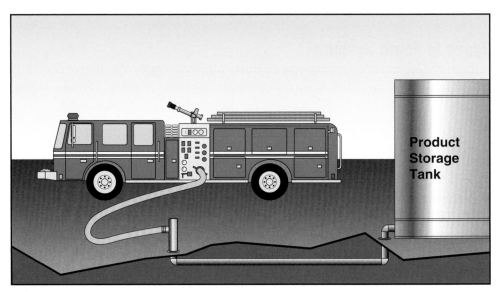

Figure 4.43 Semifixed Type A systems require foam to be supplied from a mobile fire apparatus.

Systems such as these may use subsurface injection systems, where the foam is injected at the base of a burning storage tank and allowed to surface and extinguish the fire. These systems are used where topside application may not be effective because of wind or heavy fire conditions. Because the foam solution is lighter than the product in the tank, it floats to the top when introduced at the bottom of the tank. This method is highly effective for extinguishing bulk tank fires.

Semifixed Type B

A semifixed Type B system provides a foam solution source that is piped throughout a facility-much like a water distribution system. The foam solution is delivered to foam hydrants for connection to hoselines and portable foam application devices **(Figure 4.44)**. The difference between a semifixed

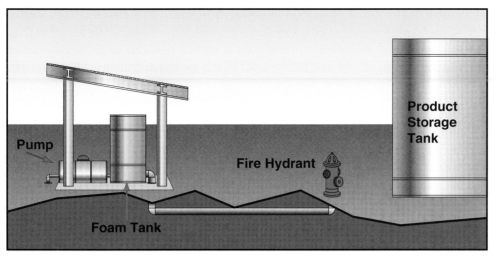

Figure 4.44 Semifixed Type B systems provide foam through a hydrant that is distributed through hoselines and portable foam devices.

Type B system and a fixed system is that a fixed system actually applies foam to the hazards, while the semifixed system merely provides foam capability to an area. In these systems, once the foam is provided to a certain location, it must then be applied manually.

High-Expansion Foam

A high-expansion foam system is designed for local application or total flooding in commercial and industrial applications **(Figure 4.45)**. These systems consist of the following primary components:

● Automatic detection or manual actuation systems

● Foam generator

● Piping from the water supply and foam concentrate storage tank to the generator

A high-expansion foam system can be actuated by any of the common fire detection devices or by a manual pull station. Electric or gasoline motors or water power the foam generators. The generators should have a fresh-air intake to make sure that foam does not become contaminated by products of combustion. In addition, venting should be provided ahead of the foam to allow it to move through the area to be protected. In the total flooding application, the building can be filled to several feet (meters) above the highest storage area or equipment within a few minutes.

Figure 4.45 High-expansion foam systems are used when total flooding applications are required, such as in aircraft hangars.

Foam/Water

A foam/water system is similar to a deluge sprinkler system but has foam capabilities. These systems are used where there is a limited foam-concentrate supply but an unlimited supply of water. As a result, if the foam-concentrate supply becomes depleted, the system will continue to operate as a straight deluge sprinkler system.

Typically, the foam/water system is an automatic system that operates in the same manner as a regular deluge system. The major difference is that the foam-induction system and special aerating sprinklers are at the end of the piping. This type of system produces a lean foam solution that eventually expands six to eight times when it is discharged from the sprinkler. This solution produces fluid foam that will flow around obstructions after it is delivered.

The foam/water system may be divided into two parts: water system and foam system. The foam system contains a concentrate tank, a pump, a metering valve, a strainer, piping, and an actuation unit. Protein, fluoroprotein (FP), and aqueous film forming foam (AFFF) concentrates may be used in foam/water systems. AFFF may also be discharged through regular water sprinklers with favorable results. When discharged through standard sprinklers, AFFF has greater velocity than when it is discharged through foam sprinklers. This type of foam tends to improve the spray and the penetration.

A foam/water system operates when the initiation devices sense the presence of fire and send an appropriate signal to the system control unit. This method, in turn, triggers the deluge valve and the system. At the same time, the deluge valve on the foam side of the system opens, the foam pump starts, and the concentrate is introduced into the waterflow. The foam/water solution flows to the sprinklers, where air is introduced to create the foam. The foam is then delivered to the target area.

After the fire is extinguished, the system must be shut down, drained, and thoroughly flushed to remove any foam residue. The concentrate tanks must then be refilled. Once this draining is complete, the valves can be reset and the system restored to service.

Foam Nozzles and Sprinklers

Foam nozzles and sprinklers, sometimes referred to as *foam makers*, are the devices that deliver the foam to the fire or spill. Fixed systems may have either handlines and foam sprinklers or generators attached to them. Standard fixed-flow or automatic water fog nozzles may be used on AFFF or film-forming fluoroprotein (FFFP) handlines. The various types of foam nozzles and sprinklers are listed as follows:

- Smooth bore nozzle
- Low-expansion foam nozzle
- Self-educting foam nozzle
- Standard fixed-flow fog nozzle
- Automatic nozzles
- Foam/water sprinkler
- High back-pressure foam aspirator
- High-expansion foam generator

Smooth Bore Nozzle

The use of smooth bore nozzles is limited to Class A, compressed-air foam system (CAFS) applications. In these applications, the smooth bore nozzle provides an effective fire stream that has maximum reach capabilities. Tests indicate that the reach of the CAFS fire stream can be more than twice the reach of a low-energy fire stream.

When using a smooth bore nozzle with a CAFS, disregard the standard rule that the discharge orifice of the nozzle should be no greater than half the diameter of the hose. Tests show that a 1½ inch (38 mm) hoseline may be equipped with a nozzle tip up to 1¼ inch (32 mm) in diameter and still provide an effective fire stream.

Low-Expansion Foam Nozzle

Foam solution must be mixed with air to form foam. The air-aspirating foam nozzle is the most effective appliance for generating low-expansion foam. The special design of a foam nozzle aerates the foam solution to provide the highest quality foam possible. Smaller foam nozzles may be handheld. Larger foam nozzles may be monitor-mounted units.

The fog/foam-type nozzle is marketed with two stream-shaping foam adapters. The basic adapter breaks the foam solution into small streams and at the same time inducts air through the venturi effect **(Figure 4.46)**. The cone-shaped attachment gives the nozzle extra reach, and the screen produces a more homogenous high-air-content foam for gentle applications **(Figure 4.47)**.

Figure 4.46 The basic adapter is attached to an adjustable fog nozzle and delivers the finished foam in small streams. *Courtesy of Tom Hughes.*

Figure 4.47 The cone adapter allows for greater reach with the foam stream and incorporates more air into the foam solution, allowing for gentle application. *Courtesy of Tom Hughes.*

Self-Educting Foam Nozzle

The self-educting foam nozzle operates with an eductor that is built into the nozzle rather than into the hoseline. As a result, its use requires the foam concentrate to be available where the nozzle is operated. The self-educting foam nozzle has the following advantages:

- It is easy to use.
- It is inexpensive.
- It works with lower pressures.
- A wide variety of flow rate versions are available.

Disadvantages include the logistical problems of relocation of the nozzle and the resulting need for relocation of the concentrate. Use of a foam nozzle also compromises firefighter safety in that personnel cannot move quickly and must leave the concentrate behind if they are required to back out of the area.

Standard Fixed-Flow Fog Nozzle

The fixed-flow, variable-pattern fog nozzle is used with foam solution to produce a low-expansion, short-lasting foam. This nozzle breaks the foam solution into tiny droplets and uses the agitation of water droplets moving through air to achieve its foaming action. The best application of this nozzle is when it is used with regular AFFF concentrate and Class A foams because its filming characteristic does not require high-quality foam to be effective. These nozzles cannot be used with protein and FP foams or any alcohol-type foams. The fixed-flow fog nozzle may be used with alcohol-resistant AFFF foams on hydrocarbon fires but should not be used on polar solvent fires.

Automatic Nozzle

An automatic nozzle operates with an eductor in the same way that a fixed-flow fog nozzle operates. However, the eductor must be operated at the inlet pressure for which it was designed, and the nozzle must be fully open. An automatic nozzle may cause problems if the eductor is operated at a lower pressure than the manufacturer recommends or if the nozzle is not fully open.

Foam/Water Sprinkler

Foam/water sprinklers are found on fixed-foam deluge and foam/water systems. These systems use AFFF and FFFP foams. Many foam/water sprinklers resemble air-aspirating nozzles in that they use the venturi effect to mix air into the foam solution. Some systems use standard sprinklers to form less-expanded foam through simple turbulence of the water droplets falling through the air. Foam/water sprinklers come in upright and pendant designs. Their deflectors must be adapted to meet the specific installation requirements.

High Back Pressure Foam Aspirator

High back pressure foam aspirators, also known as *forcing foam makers*, are used to deliver foam under pressure. These foam makers are most commonly used in subsurface injection systems that protect cone-roof hydrocarbon storage tanks, but they are also used in other applications. High back pressure aspirators supply air direction to the foam solution through Venturi action. This action typically produces low-air-content foam that is homogenous and stable.

High-Expansion Foam Generator

High-expansion foam generators produce high-air content, semistable foam. The water-aspirating type nozzle and the mechanical blower are the two basic types of high-expansion foam generators.

A water-aspirating type nozzle is similar to other foam-producing nozzles except that it is much larger and longer. The back of the nozzle is open to allow airflow. The foam solution is pumped through the nozzle in a fine spray that mixes with air to form moderate-expansion foam. The end of the nozzle has a screen or series of screens that breaks up the foam and further mixes it with air. These nozzles typically produce lower air-volume foam than mechanical blower generators.

A mechanical blower generator is similar to a smoke ejector in appearance **(Figure 4.48)**. It operates on the same principle as the water-aspirating type nozzle except that the air is forced through the foam spray instead of being pulled through by water movement. This device produces higher air content foam and is typically associated with total flooding applications.

Figure 4.48 Mechanical blower generators closely resemble smoke ejectors because they incorporate a fan system.

Testing and Inspection

Sprinkler systems require periodic inspections and maintenance in order to perform properly during a fire situation. Company managers, maintenance personnel, and fire prevention and inspection personnel should be able to inspect systems and identify problems **(Figure 4.49)**. In addition, model codes, AHJs, and insurance companies may require periodic inspections to ensure proper operation and maintenance of the system. Ultimately, the building owner is responsible for inspection, testing, and maintenance of the fire protection systems. In complex situations or those where the building owner may be absent or off-site, those responsibilities may be transferred to another qualified person or company.

NFPA® 25, *Standard for the Inspection, Testing, and Maintenance of Water-Based Fire Protection Systems*, provides the requirements for periodic testing, mainte-nance, and inspection of automatic sprinkler systems.

Figure 4.49 Inspections should be done as required by the AHJ to ensure that adequate sprinklers are available on site.

The following information introduces inspection and testing procedures but is not an all-inclusive review of the inspection process. For more information, consult IFSTA's **Fire Inspection and Code Enforcement** manual.

Acceptance Tests

In many jurisdictions, a representative from the AHJ or fire department may be required to witness sprinkler system acceptance tests. A common practice is for the representative of the installation contractor to conduct these tests. Depending on the type of system, the tests and procedures performed may include the following:

- Flushing of underground connections
- Hydrostatic tests (underground and aboveground piping)
- Waterflow alarm test
- Main drain test
- Trip test (for dry-pipe, preaction, and deluge systems)

IFSTA's **Fire Inspection and Code Enforcement** manual contains more information on acceptance testing and a variety of forms that may be used for sprinkler system inspections and testing.

Sprinklers

Sprinklers need to be inspected annually to ensure that they are appropriate for the current occupancy and in good working order. The inspection should verify that all sprinklers are clean and not painted, undamaged, and free of corrosion **(Figure 4.50)**. It should also be noted whether guards are needed to protect sprinklers against mechanical damage. Standard sprinklers in service for more than 50 years should be replaced or tested, whereas quick response sprinklers in service for more than 20 years should be replaced or tested per NFPA® 25.

Weak sprinklers may be detected by a noticeable change in the position of the fusible link or leaking around the sprinkler orifice. Sprinklers exposed to a corrosive atmosphere should have a special protective coating. Sprinklers that are corroded, painted, or loaded with foreign materials should be replaced. However, if the loading is from dust, compressed air may be used to blow the dust off of the sprinklers.

When inspecting the sprinklers, a clearance of at least 18 inches (450 mm) should be maintained under sprinklers **(Figure 4.51)**. Some sprinklers, such as large drop (CMSA) and ESFR sprinklers, require a minimum clearance of 36 inches (900 mm). For this reason, inspectors should review the code requirements for the individual occupancy. Facilities should have a supply of extra sprinklers available so that they can be replaced when a fire occurs or when an inspection determines that repairs are needed.

Inspection and Testing of Sprinkler Systems

All systems, regardless of type or occupancy, should be inspected and tested on a regular basis as required by NFPA® 25 and NFPA® 72. A written record of all tests and inspections should be made, compared to earlier tests, and kept on file for future reference.

Figure 4.50 During inspection, sprinklers are sometimes encountered that have been corroded.

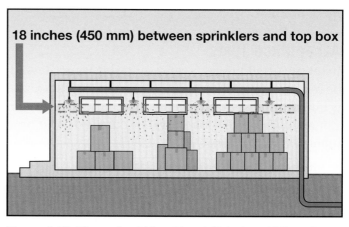

18 inches (450 mm) between sprinklers and top box

Figure 4.51 There should be at least 18 inches (450 mm) between sprinklers and stored materials.

Prior to testing and after completion, the fire department and alarm-monitoring organization (if one is used) should be notified of the alarm test. If no alarms were received (including waterflow and valve supervision), corrective maintenance service is required on the alarm system. Alarm and sprinkler contractors should coordinate this maintenance.

Wet-Pipe Systems

Inspections of wet-pipe systems are primarily concentrated in five areas: valves, sprinklers, piping, water supply, and waterflow alarms. Inspectors should ensure that the following items are in operational condition and that the indicated tests have been conducted:

- Waterflow alarm test.
- Main drain test.
- Control valve operation test.
- Valves are fully opened and secured or otherwise supervised in an approved manner.
- Valve operating wheels or cranks are in good condition.
- Valves are accessible at all times.
- Valve operating stems have not been subjected to mechanical damage.
- Operating wrenches for post indicator valves (PIVs) are in place, the "OPEN/SHUT" signs are readable, and the cover glass is clean and in place.
- PIV bolts are tight and the barrel casings are intact.
- Main drain valves, auxiliary drains, and inspector's test valves are closed.
- FDC connection threads are unobstructed and in good condition and caps are in place.

Dry-Pipe Systems

Dry-pipe systems have many elements in common with wet-pipe systems and are inspected in the same manner. However, these systems also have unique features that require special attention. During an inspection of a dry-pipe sprinkler system, inspectors should ensure that the following items are in operational condition and that the indicated tests have been conducted:

- Indicating valves are open and properly supervised in the open position.
- Air-pressure readings correspond to previously recorded readings.
- Ball-drip valves move freely and allow trapped water to seep out of the FDC.
- FDC threads are unobstructed and in good condition and caps are in place.
- The system's air pressure is at 5 to 20 psi (35 kPa to 210 kPa) (depending on the system type) above the trip point and no air leaks are indicated by a rapid or steady air loss.
- The system's air compressor is well maintained, operable, and of sufficient size.
- Waterflow alarm test.

- Main drain test.
- Control valve operation test.
- Annual dry valve trip test.
- Quarterly accelerator test.

CAUTION

System air loss is often detected by the short cycling operation of tankless air compressors or the frequent operation of tank-mounted compressors.

Deluge and Preaction Systems

Deluge and preaction systems follow the same guidelines as those performed on wet and dry systems. More frequent inspections may be needed during cold weather. The supervisory air pressure in the preaction system should be checked weekly. The detection system should be inspected based on the type of detection method employed and the requirements of NFPA® 72.

Chapter Summary

Automatic sprinkler systems and other water-based fire suppression systems are essential to life safety and property conservation because of their efficiency and reliability. Numerous lives and countless dollars have been saved since sprinkler systems were first introduced more than a century ago.

Fire service personnel must possess a good working knowledge of sprinkler systems and their components and operating methods. Inspection of systems to ensure proper operation and maintenance is an important aspect in the functioning of the automatic sprinkler system.

Review Questions

1. What are the benefits of automatic sprinkler systems?
2. What is the primary difference between residential and commercial sprinkler systems?
3. What six components are typically included in an automatic sprinkler system?
4. Define the four main characteristics of sprinkler systems.
5. Identify and describe three types of specialty sprinklers.
6. Identify the two types of wet-pipe sprinkler systems.
7. How does a dry-pipe valve operate in a dry-pipe sprinkler system?
8. Describe the three methods of operating a deluge system.
9. What are the three basic categories of preaction sprinkler systems?
10. Where are water-mist fire suppression systems found?

11. What are the four methods that foam uses to extinguish fire?

12. How is high-quality foam generated?

13. Describe the five types of foam systems.

14. What are two types of foam nozzles?

15. Depending on the type of system, what should be included when performing acceptance tests?

16. How often do sprinklers need to be inspected?

17. What items should be checked when testing and inspecting wet-pipe systems?

18. What tests should be performed when testing and inspecting dry-pipe systems?

Standpipe and Hose Systems

Chapter Contents

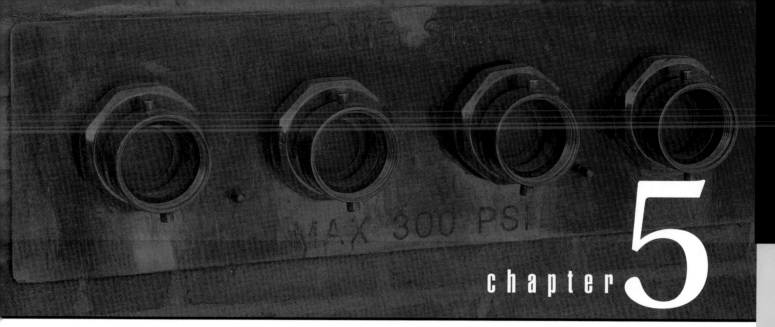

Key Terms

FESHE Outcomes

Fire and Emergency Services Higher Education (FESHE) Outcomes: Fire Protection Systems

1. Explain the benefits of fire protection systems in various types of structures.

4. Identify the different types and components of sprinkler, standpipe, and foam systems.

10. Discuss the appropriate application of fire protection systems.

Standpipe and Hose Systems

Learning Objectives

After reading this chapter, students will be able to:

1. Identify the basic components of a standpipe system.

2. Identify the different classes of standpipe systems and their intended uses.

3. Describe the different types of standpipe systems.

4. Describe the functions of a fire department connection (FDC).

5. Identify considerations concerning water supply and water pressure for standpipe and hose systems.

6. Describe the processes for initial and in-service inspections of standpipes.

7. Describe the tools and equipment fire personnel must have in order to facilitate the use of standpipe systems.

Chapter 5
Standpipe and Hose Systems

Case History

A Midwest fire department recently started a project to survey and inspect standpipes in all high-rise buildings within its jurisdiction. The project was implemented in order to determine the current status of the pressure regulating valves (PRVs) within these buildings. This department had a desire to learn from and prevent a tragic loss fire similar to those which occurred at the One Meridian Plaza in Philadelphia, the First Interstate Bank Tower in Los Angeles, or the Central Park Tower in Caracas, Venezuela.

During the first full-flow testing of the standpipe system in a 14-story building, the department quickly realized that many of the PRVs were not set correctly for fire fighting operations. The PRVs required special tools to adjust the setting on each PRV. Without this tool, static pressure on the downstream side of the PRV would have been too low to support fire fighting operations. This testing highlighted the need for the department to examine its high-rise fire fighting equipment, tactics, and strategy in addition to the fixed fire suppression systems within the buildings in their jurisdiction.

Standpipe and hose systems are designed to provide a means for rapidly deploying fire hoses and operating fire streams at locations that are remote from the fire apparatus. These systems can be found in a variety of buildings and structures where required by the authority having jurisdiction (AHJ).

The value of standpipes in expansive one-story structures is primarily one of expediency. Horizontal standpipes expedite fire control by reducing the time and effort needed to manually advance a hoseline several hundred feet (meters) to reach the seat of the fire. Horizontal standpipes can also facilitate the overhaul of fires that have been controlled by sprinkler systems by reducing the amount of hose needed to reach the area.

In many high-rise buildings, a standpipe is an essential aspect of the building's design because it is the primary means for manual extinguishment and overhaul of a fire **(Figure 5.1, p. 152)**. Because the fire pump discharge pressure necessary to reach the top floor of a 50-story building may be 350 to 400 psi (2 450 kPa to 2 800 kPa), the proper operation of the standpipe system is critical.

A **standpipe system** may be used by firefighters, properly trained occupants, or both, depending on the type of system installed. The system may be supplied by a reliable water supply and/or augmented by water from fire department

Standpipe System — Wet or dry system of pipes in a large single-story or multistory building with fire hose outlets installed in different areas or on different levels of a building to be used by firefighters and/or building occupants. The system is used to provide for quick deployment of hoselines during fire fighting operations.

Figure 5.1 Standpipes are widely used in high-rise structures to provide water for fire fighting purposes.

engines through a fire department connection (FDC) **(Figure 5.2)**. The standpipe system may also be part of or separate from an automatic sprinkler, water spray, water mist, or foam-water system.

Although standpipe systems are required in many buildings, they do not take the place of automatic sprinkler systems, nor do they lessen the need for sprinklers. Automatic sprinklers continue to be the most effective method of fire control.

Figure 5.2 Fire department apparatus can provide water and increase water pressure in the standpipe system through the fire department connection (FDC).

Components of Standpipe Systems

Components found in standpipe systems are consistent with the material descriptions and requirements outlined in Chapter 4 which include:

- Control and check valves **(Figure 5.3)**
- Pipe and fittings
- Hangers and supports
- Seismic protection
- Hose stations (the type and diameter of hose will be governed by the classification of the system)
- Water supply
- Water-flow control valves
- Risers (piping systems used to transfer water from the supply to the discharge)
- Pressure-regulating devices
- Fire department connection (FDC)

Figure 5.3 A check valve ensures the proper direction of waterflow during operations that use a fire department connection.

Classifications of Standpipe Fire Suppression Systems

NFPA® 14, *Standard for the Installation of Standpipe and Hose Systems*, is the standard for the design and installation of standpipes. This standard establishes three classes of standpipe systems. These classifications are based on the intended use of the hose stations or discharge outlets. NFPA® 13, *Standard for the Installation of Sprinkler Systems*, also contains requirements for hose stations.

Class I

Class I standpipe systems are primarily for use by fire suppression personnel trained in handling large hoselines. Class I systems must be capable of supplying effective fire streams during the more advanced stages of fire within a building. A Class I system provides 2½-inch (65 mm) hose connections or hose stations attached to the standpipe riser **(Figure 5.4, p. 154)**. The 2½-inch (65 mm) hose connections may be equipped with a **reducer** on the cap that allows for the connection of a 1½-inch (38 mm) hose coupling as well.

Class II

The Class II system is primarily designed for use by building occupants who are trained in its use or by fire department personnel. These systems are equipped with 1½ inch (38 mm) hose and nozzle and stored on a hose rack system. The hose used in these systems is typically a single-jacket type and equipped with a lightweight, twist-type shut-off nozzle. These systems are sometimes referred to as **house lines** **(Figure 5.5, p. 154)**.

There is some disagreement over the value of a Class II system. The presence of the small hose may give a false sense of security to building occupants and create the impression that they should attempt to fight a fire

Reducer — Adapter used to attach a smaller hose to a larger hose. The female end has the larger threads, while the male end has the smaller threads.

House Line — Permanently fixed, private standpipe hoseline.

Figure 5.4 A Class I standpipe system.

Figure 5.5 Class II standpipes include a hose and nozzle and are primarily designed for use by building occupants.

even though the safer course would be to escape. Fire department personnel should follow their department's standard operating procedures (SOPs) concerning the use of these systems.

Class II Hose Station

NFPA® 14 requires a 1½-inch (38 mm) hose station for use by trained personnel; however, the hose may not be provided due to the training and maintenance requirements for the hose. Another reason the hose may not be provided is the possible risk of untrained building occupants attempting to use the hose rather than exiting the building immediately. Building owners believe that supplying a hose may encourage occupants to attempt fire extinguishment even though they have not received proper training.

NFPA® 14 also requires a 1½-inch (38 mm) hose connection for fire department use. These hose connections are not always properly maintained (due to the system not being operational) and could pose a possible risk to personnel during an emergency situation. Fire department personnel should follow their department's SOPs concerning the use of these systems and should use the systems on a facility-by-facility basis.

Class III

A Class III system combines the features of Class I and Class II systems. Class III systems provide 1½-inch (38 mm) hose stations to supply water for use by building occupants who have been trained and 2½-inch (65 mm) hose connections to supply a larger volume of water for use by fire departments and those trained in handling heavy fire streams **(Figure 5.6)**. The design of the system must allow both Class I and Class II services to be used simultaneously. It is possible that the local jurisdiction may request the removal of the hose, nozzles, and rack leaving only the 2½-inch (65 mm) and 1½-inch (38 mm) discharge connections **(Figure 5.7)**.

Figure 5.6 Class III standpipes combine features of Class I and Class II systems.

Figure 5.7 Some jurisdictions have removed hose and nozzle components of Class III systems.

Types of Standpipe Systems

Within the three classes of systems, there are different types. The different types of standpipe systems include:

- **Automatic wet** — This system contains water at all times. The water supply is capable of meeting the system demand automatically. The water-supply control valve is open, and pressure is maintained in the system at all times **(Figure 5.8, p. 156)**. A wet standpipe with an automatic water supply is most desirable because water is constantly available at the hose station. Wet standpipe systems cannot be installed in a building or areas subject to freezing.

- **Automatic dry** — This system contains air under pressure to supervise the integrity of the piping. Water is admitted to the system through a dry pipe valve upon the opening of a hose valve. Automatic dry systems have a permanently attached water supply. This type of system has the disadvantage of greater cost and maintenance requirements.

- **Semiautomatic dry** — A standpipe system is permanently connected to a water supply that is provided with an inline water control device, such as a deluge valve, that requires activation of a remote manual device to release the valve and allow water to fill the standpipe. The available water volume and pressure are required to provide the minimum system demand required by NFPA® 14 without the fire department supplementing the supply at the FDC.

Figure 5.8 Automatic wet standpipe systems contain water under pressure at all times.

- **Manual dry** — This system has no water or a permanent water supply. It is designed only to transfer waterflow and pressure from a fire apparatus through the FDC to the hose connections.

- **Manual wet** — This system is frequently provided with an inadequate water supply for fire fighting activities, and supplemental water must be provided by the fire department through the FDC.

NOTE: The manual wet and automatic wet standpipes are commonly designed as a part of a sprinkler system.

Fire Department Connections

Each Class I or Class III standpipe system requires one or more FDCs through which a fire department pumping apparatus can supply water into the system. An FDC should be located in an accessible area on the exterior of the structure or away from the building (**Figures 5.9a and b**). High-rise buildings require two FDCs to supply the standpipe systems, additionally high-rise buildings having two or more zones require an FDC for each zone. A zone is a vertical subdivision of a standpipe determined by the pressure limitations of the system.

Figure 5.9a This FDC is attached to the structure.

Figure 5.9b Some FDCs are located away from the structure in a location that is easily accessible to fire department apparatus.

In high-rise buildings with multiple zones, the upper zones may be beyond the height to which a fire engine can effectively supply water. This height is usually around 450 feet (135 m), depending on the available hydrant pressure and other factors. For standpipe system zones beyond that height, an FDC is of no value unless the fire department's apparatus is equipped with a special high-pressure pump and hose, and the system has high-pressure piping.

Standard requirements specify that there shall be no shutoff valve between the FDC and the standpipe riser. In multiple-riser systems, however, gate valves are provided at the base of the individual risers to allow for maintenance while leaving the remainder of the risers operational.

Each FDC is required to have at least two 2½ inch (65 mm) connections or more as prescribed by the AHJ for fire department use **(Figure 5.10)**. In addition, there shall be at least one 2½ inch (65 mm) connection for each 250 gpm (1 000 L/min) of system demand. The hose connections on the FDC typically have a female connection with National Hose Standard (NHS) threads and should be equipped with standard cap plugs, approved breakaway covers, or locking caps **(Figure 5.11)**. If the AHJ does not use NHS-type threads, it is important that the hose-connection threads conform to those used by the local fire department. Some jurisdictions require Storz-type couplings that allow large diameter hose to supply standpipe systems.

A sign with the word STANDPIPE on a plate or fitting must indicate the type of system served by the FDC **(Figure 5.12, p. 158)**. If the FDC serves the sprinkler and standpipe systems (combination system), the sign shall state AUTO SPKR AND STANDPIPE. If the FDC does not

Figure 5.10 Some FDCs incorporate more than two 2½-inch (65 mm) connections.

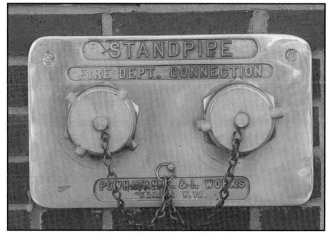

Figure 5.11 Cap plugs for FDCs are easy to remove in an emergency and prevent debris from entering the opening. *Courtesy of the McKinney (TX) Fire Department.*

service the entire building, the sign must specify which floors or areas are served **(Figure 5.13)**. When the pressure required to meet system demand is higher than what is called for in the local SOPs, the required pressure should be indicated on the sign.

Figure 5.12 FDCs that serve only the standpipe must say STANDPIPE on the identification plate.

Figure 5.13 The information plate must identify the specific areas that are served if the FDC does not serve the entire building.

Water Considerations

Two different factors to consider with regards to the use of water with standpipe systems are water supply and water pressure. These two factors must be taken into account when designing and installing standpipe and hose systems.

Water Supply

The water supply for a standpipe system may be provided by a variety of sources, including public water supplies, pressure tanks, and gravity tanks. Not all of these water sources are practical in every situation. Water supplies can be used in combination with automatic or manual fire pumps for additional pressure.

The amount of water required for a standpipe system depends on the number of standpipes within a building to a maximum required by NFPA® 14. The pressure required for a standpipe system is influenced by the number and height of standpipes.

The water supply for Class I and Class III standpipe systems should provide 500 gpm (2 000 L/min) for at least 30 minutes with a residual pressure of 100 psi (700 kPa) at the most hydraulically remote 2½ inch (65 mm) outlet. A minimum of 65 psi (455 kPa) is required for the most remote 1½ inch (38 mm) outlet.

If more than one standpipe riser is required in a building, the water supply must provide 250 gpm (1 000 L/min) for each additional riser to a maximum of 1,250 gpm (5 000 L/min) for an unsprinklered building and 1,000 gpm (4 000 L/min) for a sprinklered building. For Class II standpipes, 100 gpm (400 L/min) must be provided for at least 30 minutes with a residual pressure of at least 65 psi (455 kPa) at the most hydraulically remote outlet.

The current NFPA® 14 minimum requirement for residual pressure is 65 psi (455 kPa) for 1½-inch (38 mm) hose connections and 100 psi (700 kPa) for 2½-inch (65 mm) hose connections. However, these are minimums and may not be adequate in all instances. For example, more pressure may be needed to supply a fog nozzle on the end of a 100-foot (30 m) hose connected to the topmost hose outlet. In these situations, local codes may require higher minimum residual pressures. The code adopted by the local AHJ should be consulted to determine the minimum residual pressures for standpipe-protected occupancies.

Minimum Required Pressure

Current codes require a minimum pressure of 100 psi (700 kPa) at the most remote Class I or III fire hose valve. The minimum required pressure provided does not take into consideration changing fire department tactics utilizing automatic nozzles and smaller diameter hoselines, as previous code editions required a minimum of 65 psi (455 kPa) which was based on a 50 psi (350 kPa) smooth bore nozzle pressure and 15 psi (105 kPa) friction loss for 2½-inch (65 mm) diameter hose. Fire departments should assess the minimum operating pressure of nozzles and friction loss associated with standard fire hose lengths in order to supplement the minimum required fire fighting demand at the FDC.

High-Rise Buildings

The height of the building and the standpipe service class determine the size of the standpipe riser. For Class I and Class III service, the minimum riser is 6 inches (150 mm) unless hydraulic calculations are performed justifying a 4-inch (100 mm) riser. Standpipes that are part of a combined system (those that include a sprinkler and standpipe) must be at least 6 inches (150 mm) in diameter. NFPA® 14 requires that all standpipes be calculated via the automatic water source or FDC to verify minimum requirements. For Class II service, a standpipe should not be less than the minimum diameter as permitted by NFPA® 14.

NFPA® 14 only requires a standpipe in each required exit stairway, with additional hose stations provided such that no area is more than 200 feet (60 m) from a hose connection in sprinklered buildings or 150 feet (45 m) in unsprinklered buildings. Building codes, such as the IBC, require standpipes located within horizontal exits or exit passage ways so that any part of a floor is within 130 feet (40 m) of the standpipe hose connection. In addition, standpipe fire hose valves are required within covered mall buildings and roofs that have a slope of less than 33 percent.

This distance allows any fire to be reached with 100 feet (30 m) of hose, plus a 30-foot (9 m) fire stream. Standpipes and their connections are most commonly located within noncombustible fire-rated stair enclosures so that firefighters have a protected point from which to begin an attack. If the building is so large that the standpipes located in the stairwells cannot provide coverage to the entire floor, additional stations or risers must be provided. Some buildings with rated horizontal exits will have a fire hose connection on each side of the opening.

NOTE: Recent changes in codes and standards now require fire hose valves to be located on the intermediate level within stairwells and not on the floor landing. This change is impacting fire department SOPs when requirements are to connect to the floor below the incident.

The standpipe hose connections should be located not less than 3 feet (0.9 m) and not more than 5 feet (1.5 m) from floor level. These connections should be plainly visible and should not be obstructed. Any caps over the connections should be easy to remove.

Buildings equipped with a Class I or Class III system may be required to have a 2½ inch (65 mm) outlet on the roof. This outlet may be required when the building has any of the following present:

● A combustible roof

● A combustible structure or equipment on the roof

● Exposures that present a fire hazard

● A roof garden or landscaped roof

Water Pressure

One problem encountered with standpipes installed in high-rise buildings is the higher pressures encountered in the lower portions of the system. This is a result of the height of the building. It is important to remember that 0.434 psi (3 kPa) is required to raise water 1 foot (0.3 m or 300 mm). When a building is 10- or 20-stories tall, the pressure on the system at the lower floors is so great that it makes the hose difficult to handle when it is attached to the hose stations on those floors. In these cases, pressure-reducing and/or pressure control valves may be required to keep operating pressures below 175 psi (1 200 kPa) **(Figure 5.14)**.

High-rise buildings may have several zones **(Figure 5.15)**. For example, the Willis Tower (formerly Sears Tower) in Chicago is 1,400 feet (425 m) tall and has a standpipe system divided into seven zones. Water for the upper zones is supplied by an individual zone fire pump that is supplied directly from a water main, or it may receive water from the fire pump that supplies the lower zone.

In high-rise buildings with several zones, the upper zone pumps may be arranged to draft from tanks on the upper floors. The tanks, which hold several thousand gallons (liters) of water, are filled automatically from lower-zone fire pumps and/or the domestic water-supply pumps by means of automatic float valves. These float valves open when the level in the tank begins to drop. Tanks located on the upper floors for a source of water supply provide for increased system reliability. Should the lower zone pumps fail, there is still water available to the pumps located on the upper floors of the building.

Figure 5.14 Pressure reducing valves are sometimes necessary when standpipes are used on lower floors in order to reduce pressure at the outlet to a more manageable level.

Figure 5.15 Multiple zones may exist in larger buildings.

NFPA® standards require a pressure-regulating device at a hose outlet that exceeds 100 psi (700 kPa) for a 1½ inch (38 mm) connection and 175 psi (1 200 kPa) for a 2½ inch (65 mm) connection. This device will limit the pressure to 100 psi (700 kPa) unless the fire department has approved otherwise.

Pressure-regulating devices prevent pressures that make hoses difficult or dangerous to handle. They also enhance system reliability because individual zones are extended to greater heights. Although this may make system design more complex, system economy can be improved by eliminating the number of pumps needed.

Three basic categories of pressure-regulating devices include the following:

- **Pressure-restricting devices** — Consist of a restricting device that allows the static inlet and outlet pressures to remain the same with the valve open. The amount of pressure drop through the restriction depends on the orifice diameter or the maximum opening of the control valve and available flow and pressure within the system. Each standpipe discharge connection is fitted with a restricting device with either different orifice sizes or adjustments to the fire hose valve for each floor and application. Pressure-restricting devices are limited to systems with 1½ inch (38 mm) hose discharges and 175 psi (1 200 kPa) maximum pressure. These devices are not a preferred type because they do not control or reduce the static water pressure in the system.

- **Pressure-control valves** — Pilot-control system senses downstream pressure and automatically opens or closes the valve to maintain the desired pressure setting.

- **Pressure-reducing valves** — Fire hose valves that reduce the static and residual pressures at the outlet that can either be factory set or field adjustable. For managing excessive pressure in excess of 175 psi (1 200 kPa) and uses a mechanism that compensates for variations in pressure. These mechanisms balance the available pressure within the system with the pressure required for hoseline use. Some pressure-reducing valves are field-adjustable when a situation arises where too little pressure is being supplied.

If a pressure-regulating device is not properly installed or is not properly adjusted for the required inlet pressure, outlet pressure, and flow, the available flow may be greatly reduced and fire fighting capabilities seriously impaired. Manufacturer instructions for installation, adjustment, testing, and maintenance must be followed carefully.

Standpipe Inspection and Testing

In order to ensure both compliance with local codes and standpipe operability, standpipes should be inspected when they are first installed and periodically thereafter. In addition to NFPA® 14 and 25, consult local codes and ordinances regarding standpipe system requirements and installations. The following sections highlight inspection and testing procedures for standpipes.

Initial Installation Inspection and Tests

A standpipe system is a significant component in a building's design. Before system installation, detailed design plans should be submitted to the AHJ for approval. It is during this phase where the plans are checked for compliance with the jurisdiction's current codes. As construction proceeds, the installation of the standpipe will need to be evaluated. In high-rise buildings, the standpipe should be in partial operation as construction proceeds to provide protection during construction.

When installation is complete, the following tests and inspections shall be performed:

- Hydrostatically test the system at a pressure of at least 200 psi (1 400 kPa) for 2 hours to ensure tightness and integrity of fittings. If the normal operating pressure is greater than 150 psi (1 050 kPa), test the system at 50 psi (350 kPa) greater than its designed pressure.

- Flush the underground piping supplying the standpipe system to remove any construction debris and to ensure that there are no obstructions.

- Perform a flow test on systems equipped with an automatic fire pump. Do this at the highest outlet to ensure that the fire pump will start when the hose valve is opened.

- Test the system to ensure that the required flow and pressure can be provided at the highest fire hose valve.

- Inspect all devices to ensure that they are listed by a nationally recognized testing laboratory such as Underwriters Laboratory Inc. (UL) or FM Global.

- Check hose stations and connections to ensure that they are located not less than 3 feet (0.9 m) and not greater than 5 feet (1.5 m) from the floor and are positioned so that the hose can be attached to the valve without kinking.

- Test pressure regulating valves.

- Test dry pipe valves and deluge valves.
- Conduct a main drain test.
- Test alarm and supervisory devices.
- Inspect each hose cabinet or closet for a conspicuous sign that reads FIRE HOSE and/or FIRE HOSE FOR USE BY OCCUPANTS OF BUILDING.
- Check FDCs for the proper fire department thread and proper signage.
- Check for a sign indicating MANUAL STANDPIPE FOR FIRE DEPARTMENT USE ONLY when a manual standpipe is installed.

CAUTION

Take extra care during testing with flow drainage and discharge. This water is often extremely dirty and can easily stain exterior building finishes.

In-Service Inspections

As with all fire protection systems, standpipe and hose systems need to be inspected and tested at regular intervals. Building management should perform a visual inspection of the standpipe and hose systems at least monthly.

NOTE: A fire protection contractor or the building operating staff members (if they have sufficient knowledge) should conduct actual testing of standpipes.

Fire department personnel should also regularly inspect standpipe and hose systems because fire fighting operations are often dependent on their functionality. During these fire department inspections, personnel should ensure that the following conditions are met:

- All water supply valves are sealed in the open position.
- All alarms, supervisory devices, dry pipe valves, preaction valves, and pressure regulating devices have been tested within the frequency prescribed by NFPA® 25 and the manufacturers' recommendations.
- Power is available to the fire pump.
- The fire pump has a current inspection tag and has no noted deficiencies.
- Individual hose valves are free of paint, corrosion, and other impediments.
- Hose valve threads are not damaged and match fire department couplings **(Figure 5.16, p. 164)**.
- Hose valves face a direction that will allow the easy attachment of hose and will prevent kinks in hose.
- FDC and caps are in place and any locking caps can be opened **(Figure 5.17, p. 164)**.
- Pipes are free of trash or debris.
- Hose valve wheels are present and not damaged.
- Hose cabinets are accessible.

- Hose is in good condition, has proper dryness, and is properly positioned on the rack.
- Hose nozzles are present and in good working order.
- Dry standpipes are drained of moisture.
- Access to the FDC and closest hydrant is not blocked.
- FDC is free of obstruction and the swivels rotate freely.
- Water-supply tanks are at the proper level.
- Hose valves on dry systems are closed.
- Dry systems are hydrostatically tested every five years.

Figure 5.16 Hose threads should be inspected to ensure that they are not damaged and that they match threads used by the fire department.

Figure 5.17 Locking FDC caps should be tested to ensure they can open easily.

Fire Department Operations

For standpipe systems to be effective, fire department personnel must be trained in their operation and understand the systems and their limitations. Most fire departments perform preplanning activities. As a part of this process, firefighters should locate the FDC, water supply valves, pumps, tanks, and hose connections. Fire departments should also develop SOPs for the use of standpipe and hose systems.

Emergency responders must have the proper equipment to facilitate the use of the standpipe systems. Most departments have standpipe packs or high-rise packs on their apparatus **(Figure 5.18)**. Regardless of their name, these packs contain the equipment necessary to connect to the standpipe system. The equipment carried in these equipment packs may vary from jurisdiction to jurisdiction, but should include 100 to 200 feet (30 to 60 m) of hose, a nozzle, and miscellaneous tools and wrenches.

The first fire fighting crew on the scene of a building with a standpipe system should take this pack into the building when investigating the alarm or call. The next due engine should stand by or connect to the FDC to supply the system. When charging the standpipe system, the jurisdiction's SOPs must be followed. Firefighters should use a hose connection that is located in a protected stairway enclosure if possible as opposed to one that is located in a corridor or open area.

Chapter Summary

Standpipes and hose systems reduce the time and effort needed by fire fighting personnel to attach hoselines and apply water during a fire emergency. In large or very tall buildings, more effort and resources would be required to stretch hoselines if standpipes or FDCs were not already in place. Because it is easy to take built-in systems for granted, fire fighting personnel must ensure that fire suppression systems are inspected and in good working order. There is no time to repair a system during an emergency.

Review Questions

1. What are the components of a standpipe system?

2. Differentiate between the three classes of standpipe systems.

3. Identify the five types of standpipe systems.

4. What is the purpose of a fire department connection (FDC)?

5. What factors influence the water supply requirements for standpipes?

6. What are the location requirements of standpipes in high-rise buildings?

7. What are three types of pressure-regulating devices?

8. What tests should be performed on a standpipe system after its installation is complete?

9. What are ten items that fire departments should regularly inspect on standpipe systems during in-service inspections?

10. What is the purpose of a standpipe pack or high-rise pack?

Figure 5.18 High-rise packs include the equipment necessary to deploy a hoseline from a standpipe and are readily available on the fire apparatus.

Fire Pumps

Chapter Contents

Key Terms

FESHE Outcomes

Fire and Emergency Services Higher Education (FESHE) Outcomes: Fire Protection Systems

4. Identify the different types and components of sprinkler, standpipe, and foam systems.

10. Discuss the appropriate application of fire protection systems.

Fire Pumps

After reading this chapter, students will be able to:

1. Describe the roles of fire pumps in fire suppression systems.

2. Identify types of fire pumps.

3. Describe types of pump drivers.

4. Describe the operation of pump controllers.

5. Identify the basic components and accessories common to all fire pump installations.

6. Describe location considerations and component arrangements for the installation of a fire pump.

7. Summarize primary performance criteria for fire pump testing.

8. Summarize inspection and maintenance procedures for fire pumps.

Chapter 6
Fire Pumps

Case History

A fire pump was installed in a university research building to provide an adequate water supply for the sprinkler and standpipe system. The fire pump was provided with a flowmeter in a bypass line around the pump but without a test header to permit discharging water during an annual pump test. For the first few years after installation, the annual pump test was conducted using the flowmeter and the pump test results were satisfactory. However, NFPA® 25, *Standard for the Inspection, Testing, and Maintenance of Water-Based Fire Protection Systems*, requires that a test in which water is discharged through hoses be conducted once every three years. Conducting such a test was a problem in this building because no test header was provided. The problem was solved by connecting hoses to the hose connections on one of the standpipes and running the hoses down the stairwell to the outside. The hoses were attached to Underwriters playpipes to measure the flow. By conducting the test in this manner, it was discovered that the pump, rated at 1,250 gpm (5 000 L/min), could only deliver about 900 gpm (3 600 L/min), which was a serious deficiency. It was ultimately determined that the poor performance was not due to the pump. It was caused by a defective backflow preventer on the suction side of the pump, which was only partially opening. This experience clearly demonstrated the desirability of a pump test header and the importance of actually flowing water through hoselines and not always depending upon testing using only a flowmeter in a bypass.

The Role of Fire Pumps in a Fire Suppression System

Many fire suppression systems would be unable to sufficiently perform the job of fire extinguishment or containment without fire pumps. Fire pumps, where required as a component of a fire suppression system, play a critical role in the integrity of the system. To achieve this role, fire pumps:

- Increase the water supply pressure available from public/private water systems and gravity tanks

- Transfer water from reservoirs, lakes, ponds, swimming pools, or other such nonpotable static water sources

- Provide adequate water volume and pressure to private fire hydrants and/or water-based fire suppression systems

A **fire pump** is a fixed pump that supplies water to a fire suppression system. The main function of the pump is to increase the pressure of the water that flows through it. Usually a fire pump is needed to supply a sprinkler or standpipe system because the available water supply source cannot supply the needed pressure. Fire pumps should be installed within a two-hour rated enclosure so that they can function as necessary in the event of a fire.

The earliest fire pumps were positive-displacement types, using rotary gears or pistons. Modern fire pumps can be either positive-displacement or centrifugal types. A positive-displacement pump works on the principle that when pressure is applied to a confined liquid, the same pressure is equally transmitted in all directions. It captures a specific volume of fluid per pump revolution. Positive-displacement pumps may be rotary lobe, rotary vane, or piston plunger.

The most common type of fire pump is the centrifugal type **(Figure 6.1)**. **Centrifugal pumps** are those that use an impeller or turbine to create pressure by the action of centrifugal force. These pumps may be installed in either a horizontal or vertical case.

Figure 6.1 The centrifugal type fire pump is most commonly used in fire protection applications.

Every pump must have a driver to operate the pump and a controller to control the driver. Additionally, every fire pump must be installed in a system of piping and valves. This chapter addresses the types of fire pumps commonly installed in commercial, residential, and industrial facilities to support existing water supplies. The pump components and accessories, common fire pump types, pump drivers, and controllers are also explained in this chapter. The chapter also presents information on testing, inspection, and maintenance of fire pumps. Criteria for the installation and acceptance testing of fire pumps are listed in NFPA® 20, *Standard for the Installation of Stationary Pumps for Fire Protection*. Criteria for the inspection, testing, and maintenance of fire pumps can be found in NFPA® 25, *Standard for the Inspection, Testing, and Maintenance of Water-Based Fire Protection Systems*.

Common Types of Fire Pumps

To support existing water supplies, three types of fire pumps are commonly installed in all occupancy types:

- **Horizontal split-case centrifugal pump** — The casing for the shaft and impeller is split in the middle and can be separated. This exposes the shaft, bearing, and impeller and makes for easy access for repair or replacement of parts **(Figure 6.2)**.

- **Vertically mounted split-case centrifugal pump** — Split-case pumps can be mounted vertically or horizontally.

- **Vertical-shaft turbine pump** — Used when the water supply is from a non-pressurized source such as a well, pond, river, or underground storage tank.

These types of fire pumps are explained in the following sections. Two additional types of pumps to be mentioned include the end-suction pump and the vertical-inline pump.

Figure 6.2 A horizontal split-case centrifugal pump.

Horizontal Split-Case Centrifugal Pumps

The **horizontal split-case centrifugal pump** is the most common of the split-case types and is acceptable for use where water can be supplied to the pump under some pressure. This pump is not self-priming, so it must be supplied with a positive suction pressure from either a public/private water supply system or an elevated tank. A pump that needs **priming** is unacceptable where automatic pump operation is required. Other features of the horizontal split-case centrifugal pump include:

- The shaft is oriented horizontally, as its name implies.
- A grease fitting is provided for the bearing at each end of the shaft for proper lubrication at regular intervals.
- Fiber packings are located on each end of the shaft in the packing gland **(Figure 6.3, p.172)**.
- Fiber packings seal the shaft and prevent excess water leakage around the shaft. They are water cooled and lubricated so that a small amount of leakage is required. Too much or too little leakage requires tightening or loosening of the packing gland **(Figure 6.4, p. 172)**.

In the horizontal split-case centrifugal pump, the water pressure is increased due to the operation of a rotor, called an **impeller**, inside the pump casing. Water is fed into the center, or eye, of the impeller from one or both sides and then thrown to the outer edges by the rotation of the impeller **(Figure 6.5, p. 172)**. As the water is forced to the outer edge of the impeller, its pressure is increased depending on how fast the impeller is turning and the diameter and design of the impeller. The impeller turns on a shaft that is usually driven by an electric motor or a diesel engine. Though relatively rare, pumps driven by a steam turbine may be encountered. The sequence of waterflow through a centrifugal pump can be seen in **Figure 6.6, p. 173**.

Horizontal Split-Case Centrifugal Pump — Centrifugal pump with the impeller shaft installed horizontally and often referred to as a split-case pump. This is because the case in which the shaft and impeller rotates is split in the middle and can be separated, exposing the shaft, bearings, and impeller.

Priming — To create a vacuum in a pump by removing air from the pump housing and intake hose in preparation to permit the drafting of water.

Impeller — Vaned, circulating member of the centrifugal pump that transmits motion to the water.

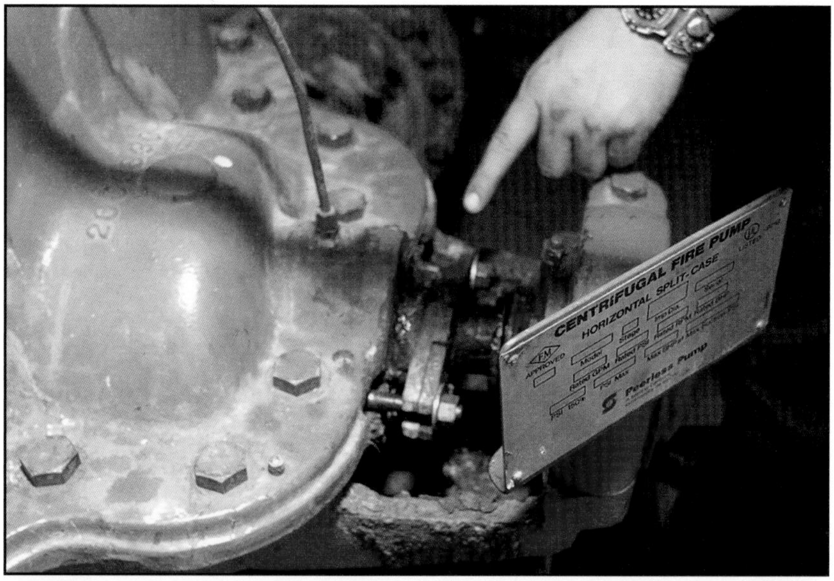

Figure 6.3 Packing glands are designed to seal the shaft and prevent excess water leakage.

Figure 6.4 Packing glands may need to be loosened or tightened to obtain the correct leakage.

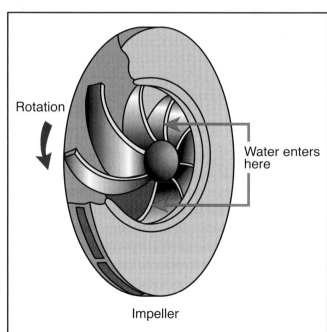

Figure 6.5 Water enters the impeller at the center and is forced to the edges by rotation.

Rotation

Water enters here

Impeller

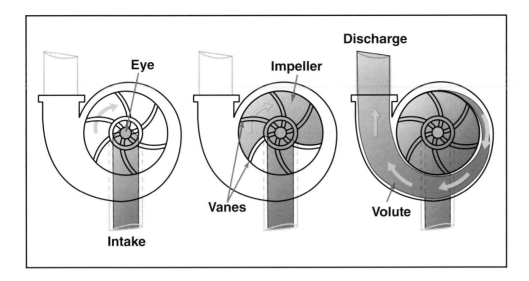

Figure 6.6 This illustration shows the path that water takes in a centrifugal pump.

Fire pumps are referred to as single-stage or multiple-stage. A **single-stage pump** is one with a single impeller. Most horizontal split-case centrifugal pumps are single-stage pumps. **Multiple-stage pumps** have more than one impeller and can deliver higher pressures and are commonly used to supply water to standpipes in high-rise buildings.

Fire pumps are listed with ratings as small as 25 gpm (100 L/min) and as large as 5,000 gpm (20 000 L/min) or greater. There is no standard pressure rating for single-stage horizontal split-case centrifugal pumps, and they may be purchased with pressure ratings varying from 40 psi (280 kPa) to 290 psi (2 000 kPa) or higher. Standard fire pumps are not permitted to have a pressure rating less than 40 psi (280 kPa). However, where the combined discharge pressure at the fire pump exceeds 175 psi (1 200 kPa), extra heavy fittings, pressure reducing valves, and other safety precautions would be required by NFPA® Standards 13, 14, and 20.

Vertically Mounted Split-Case Centrifugal Pumps

Even though the name horizontal split-case centrifugal pump implies that the pump will be installed with the drive shaft in the horizontal position, there is a modern variation. In some installations where the pump is driven by an electric motor, the motor may be installed on top of the pump. This is a standard installation technique that saves floor space. In these instances, the type of pump is referred to as a **vertically mounted split-case centrifugal pump**.

The vertically-mounted pump is basically the same as the horizontal split-case centrifugal pump and has identical applications. The impeller rotates on a shaft within the pump casing. The water is discharged in a direction perpendicular to the shaft, and the shaft rotates on ball-type bearings at each end.

Vertical-Shaft Turbine Pumps

The vertical-shaft turbine pump was originally designed to pump water from wells. This type of pump is still used where the water supply is from a nonpressurized source, including wells, ponds, rivers, and underground storage tanks **(Figure 6.7, p. 174)**. Vertical-shaft pumps never require priming because the rotating turbines are positioned inside the water source. These pumps are multistage pumps. The number of impellers is determined by the desired

Single-Stage Pump — Centrifugal pump with only one impeller.

Multiple-Stage Pumps — Any centrifugal fire pump having more than one impeller.

Vertically Mounted Split-Case Centrifugal Pump — Centrifugal pump similar to the horizontal split-case, except that the shaft is oriented vertically and the driver is mounted on top of the pump.

pressure rating — the more impellers there are, the greater the pressure that will be developed. As the water exits one impeller, it enters the next, and so on until it is discharged into the fire suppression system piping **(Figure 6.8)**.

The shaft can either be water or oil lubricated. A strainer is located at the bottom of the shaft to keep fish, snails, leaves, and other objects out of the pump. Each turbine sits in a slight enlargement of the casing that is referred to as the *bowl*. The bowl has a wear ring to prevent deterioration of the bowl itself.

An electric motor, a steam turbine, or a diesel engine turns the shaft. Fiber packing is at the top of the shaft to seal the shaft against water leakage. The fiber packing is contained inside a packing gland, which can be either tightened or loosened to obtain the proper packing lubrication.

The vertical-shaft turbine pump has been used to support pressurized water supplies. In this application it is commonly referred to as a *can-type installation*. Essentially, a cylinder or canister is

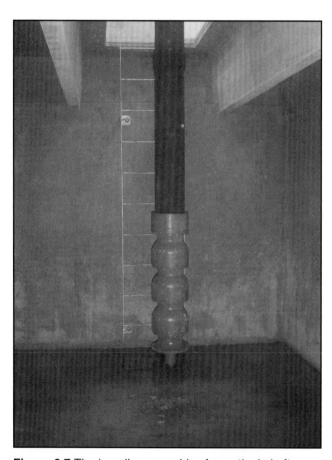

Figure 6.7 The impeller assembly of a vertical-shaft turbine pump is positioned inside the water source. *Courtesy of Floyd Luinstra.*

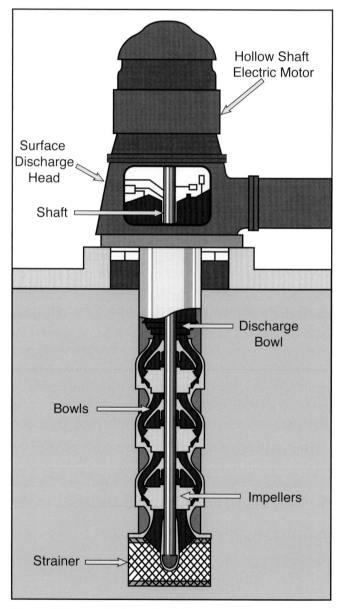

Figure 6.8 In vertical-shaft turbine applications, water moves from one impeller to the next until it is discharged from the pump.

constructed with the municipal water supply feeding this canister. The vertical-shaft pump is simply set into this canister and is able to boost the pressure of the incoming water supply.

Even though the impellers of a vertical-shaft pump are located belowground inside the water source, the driver and control panel will be accessible aboveground **(Figure 6.9)**. Vertical-shaft pumps are usually driven by an electric hollow-shaft motor mounted above the pump or by a diesel engine through a right-angle gear drive.

End-Suction Centrifugal Pumps

The end-suction centrifugal pump is a variation of the horizontal split-case pump design. End suction centrifugal pumps are single-stage pumps that have centerline suctions and discharges. These pumps differ from split-case centrifugal pumps in that the suction line is piped directly to the eye of the impeller. These pumps have pressure ratings from 40 to 150 psi (280 kPa to 1 050 kPa), along with flow ranges of 50 to 750 gpm (200 L/min to 3 000 L/min).

Advantages of the end suction centrifugal pump are the ease of installation, simplified piping arrangement, and reduced pipe strain. The pumps are self-venting, which eliminates the need for an automatic air-release valve.

Vertical-Inline Centrifugal Pumps

The vertical-inline centrifugal pump is a single-stage pump designed to fit into the intake/discharge line with the driver located above the inline impeller. The vertical inline centrifugal pump is very similar to the vertically mounted split-case centrifugal pump except that the vertical inline pump requires less floor space. The advantages of a vertical inline centrifugal pump are the ease of installation as a replacement pump, the small space required for the pump, and the ease of maintenance of the pump and driver. Vertical inline centrifugal pumps have a capacity up to 1,500 gpm (6 000 L/min) and operating pressures up to 165 psi (1 150 kPa).

Figure 6.9 The driver and control panel of a vertical-shaft pump are accessible aboveground.

Pump Drivers

The **driver** is a vital component of a fire pump installation. Drivers are the engines or motors used to turn the pump. The two common types of drivers that are presently acceptable for use with fire pumps are electric motors and diesel engines. Steam-driven pumps, although acceptable by NFPA® 20, are not commonly found. Some earlier types of engines, such as gasoline, natural gas, and propane, can still be found in older installations, but are no longer permitted for new installations.

The pump driver must have enough power to turn the pump at rated speed under all required load conditions, which include pumping at **churn** and pumping at 150 percent of rated capacity. A pump is said to be operating at churn or shutoff when it is running but all discharges are closed. The horse-

Driver — Engine or motor used to turn a pump.

Churn — Rotation of a centrifugal pump impeller when no discharge ports are open so that no water flows through the pump.

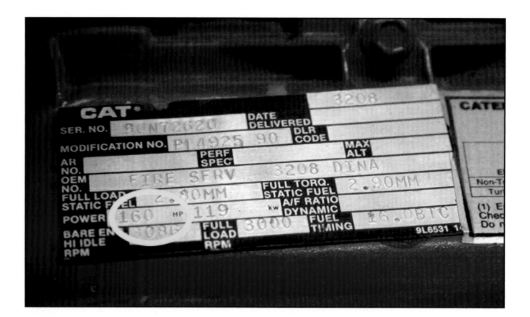

Figure 6.10 Horsepower ratings are typically found on an information plate attached to the pump driver.

power (hp) ratings are commonly found on an information plate located on the pump driver **(Figure 6.10)**. Horsepower varies according to the speed at which the driver is operating. For this reason, it is important to keep a pump driver well maintained and properly adjusted. Otherwise, the pump cannot be expected to meet its performance specifications.

Electric Motors

Electric motors have long been a dependable source of power for driving centrifugal fire pumps **(Figure 6.11)**. The motor must have adequate power output (horsepower [hp], [kilowatt (kW)]) to drive the fire pump. The required pump horsepower is determined by the pump capacity (gpm), the net pressure (discharge pressure minus the incoming pressure), and the pump efficiency. For a 1,000 gpm (4 000 L/min) pump rated at 100 psi (700 kPa), a motor of about 80 hp (60 kW) would be needed.

Electric motors powerful enough to power fire pumps use a great deal of electricity and may require a larger electrical service to the building than would be needed otherwise. Most installations also require the addition of an electrical feed from the transformer directly to the fire pump controller. The electrical feed to the fire pump motor must originate upstream of any shut-off mechanisms installed in the building, so that if the power is shut off to the building during a fire event, the fire pump can continue to run. For some types of pumps, such as the horizontal split-case centrifugal pump, the electric motor is located on a framework beside the pump to ensure proper shaft alignment. As mentioned earlier, some split-case centrifugal pumps are installed vertically and have the motor located on the top to save floor space.

A substantial advantage of the electric motor is the relatively small amount of maintenance required. While electric motors are easy to maintain, they are often neglected for this very reason. Proper lubrication of the motor bearings in accordance with manufacturer's instructions is critical for reliable performance of the motor. In addition, electric pumps should be operated weekly or monthly as required by NFPA® 25 and permitted to run for at least ten minutes to ensure they are operating properly.

A disadvantage of the electric motor involves reliability. Storms, fires, or other accidents involving power lines, transformers, or substations can leave motors without power and fire pumps useless. Wiring installations providing power for the electric motors and the controllers are required to comply with the provisions of NFPA® 70, *National Electrical Code®*.

Each electric motor should have an information plate that provides the current and voltage ratings, hp, revolutions per minute (rpm), and service factor. These data are useful in determining the acceptability of pump performance during the testing procedures.

Figure 6.11 Electric motors are a popular choice for pump drivers.

Diesel Engines

The diesel engine is a common and reliable means of powering fire pumps. Although diesel engines are usually more expensive than electric drivers, they may be a better choice because they do not rely on external power. While electrically driven pumps are simpler and require less maintenance, the diesel engine has proven to be the most dependable of all the internal combustion engines and is currently the only kind of internal combustion engine considered acceptable for fire protection applications.

Diesel engines are listed by UL and approved by FM Global if they are to be used for fire protection applications. This means that not all diesel engines are acceptable for driving fire pumps. If a diesel engine driver is used on a fire-pump application, the testing agencies

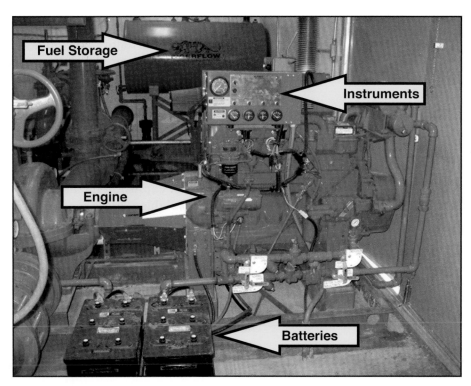

Figure 6.12 Diesel engines are currently the only kind of internal combustion engine considered acceptable for fire protection applications. *Courtesy of Sand Springs (OK) Fire Department.*

require that the engine be equipped with overspeed shutdown devices, tachometers, oil pressure gauges, and temperature gauges. The following aspects of a diesel engine must be considered when looking at its applicability to fire-pump use: engine power, engine requirements, cooling, and fuel storage **(Figure 6.12)**.

Engine Power

If the engine operates at speeds that are too high, the pump can develop excessive water pressure that can damage both the engine and the pump. If the engine operates too slowly, the pump will not develop rated pressures. The engine is therefore required to have an adjustable **governor** to maintain engine speed within a 10-percent range. The governor is set to maintain rated-pump speed at maximum pump load. If the engine speed exceeds 20 percent of the rated-engine speed, the governor will shut down the pump. The shutdown device is required to send a trouble signal to the control panel until it is manually reset.

Engine Requirements

Several instruments must be placed on a panel that is securely fastened to the engine at an accessible location. NFPA® 37 requires the following instruments:

- Tachometer — Indicates rpm of the engine
- Oil pressure gauge — Indicates the pressure of the lubricating oil
- Water temperature gauge — Indicates the temperature of the water in the engine cooling system

Engines may be started pneumatically (compressed air), hydraulically (pressurized fluids), or electrically (by storage batteries). Batteries are the most common means of starting diesel engines for fire pumps. If batteries are used, two means must be provided for recharging. One method may be a generator or an alternator that comes with the engine, and another possibility is an automatic charger. The charger must be incorporated into the design of the controller and must be capable of fully recharging the batteries within 24 hours. Place the batteries so that they are not subject to flooding, mechanical damage, extreme temperature variations, or vibration. The batteries should also be readily accessible for easy servicing.

If a diesel engine is located in an environment that is subjected to flammable vapors, starting the engine through the ordinary electric starting means could create a fire or explosion hazard. For this reason, pneumatic and hydraulic starting is available. This is accomplished by forcing a compressed gas or water through a turbine that turns the engine. For safety reasons, diesel engine exhaust fumes should be piped to the outside of the building. Do not locate exhaust piping close to combustible materials. Inspect the exhaust system regularly for leaks.

Cooling System

Diesel engines are water-cooled and make use of a closed-circuit-type cooling system. The basic components of the system include an engine driven water pump, a heat exchanger, and a reliable device for regulating the water temperature in the engine jacket. The heat exchanger works by taking water from the discharge side of the pump to cool the water in the engine. The exchange of heat from the engine water to the pump discharge water takes place in the heat exchanger. The provision of water from the discharge side of the pump requires a special piping arrangement, including a bypass line, valves, pressure regulator, and strainers.

Fuel Storage

Diesel fuel is not as volatile as gasoline, but it can still be dangerous and must be handled carefully. Safe storage, transmission, and adequate quantities must be provided. Any exposed fuel lines must be protected against mechanical damage. Fuel piping should be rigid except where the fuel line connects to the engine. At that point, flame-resistant flexible hose is required.

Example

Appropriately sized diesel fuel storage tanks should provide 1 gallon of fuel per horsepower plus 5 percent for expansion and 5 percent for sump.

- *Customary* — A pump driven by a 200 hp diesel engine would require a 220 gallon tank (200 gallons + 10 gallons for expansion + 10 gallons for sump).

- *Metric* — A pump driven by a 200 hp diesel engine would require an 880 liter tank (800 liters + 40 liters for expansion + 40 liters for sump).

To prevent runoff from any leakage from the tank, containment should be provided for environmental protection by construction of a small dike around the base of the tank. Codes and ordinances usually require the containment volume to be at least as large as the total volume of the tank. Other containment methods include sloping the floor to channel leaks to a containment basin or placing a sill around the pump room. There are numerous other considerations with regard to system design that are outside the scope of this manual. For more information, please refer to NFPA® 20.

Pump Controllers

A **pump controller** is another critical aspect of a fire pump installation. Pump controllers are the control panels used to switch the pumps on and off and control their operation **(Figure 6.13, p. 180)**. These panels govern the operation of the pump. The controller can be designed to operate the pump automatically by use of microprocessors, simple electronic circuits, or manual operation **(Figure 6.14, p. 180)**. Electric pump controllers may or may not have a power cutoff between the power company connection and the controller. Therefore, only qualified personnel should service the pump controller because of the high-voltage wiring it contains. The following sections address the controllers for electric motor-driven pumps and diesel engine-driven pumps.

Pump Controller — Electric control panel used to switch a fire pump on and off and to control its operation.

Controllers for Electric Motor-Driven Pumps

One of the nationally recognized testing laboratories should test the controller for an electrically driven pump and list for fire protection use. The controller should be located inside the pump room and as close to the pump as possible. The controller should be protected from water discharge, and all current-carrying parts of the controller should be at least 12 inches (300 mm) off the floor. The main parts of the controller for electric drivers are as follows, and each part can be found by associated number in **Figure 6.15, p. 181**:

1. **Circuit breaker** — Provides overcurrent protection by opening if too many amperes (amps) are being drawn. The circuit breaker should be accessible and operable from the outside of the controller. Proper load rating of the circuit breaker is important. It should allow at least 114 percent of the rated full-load current without tripping and also permit normal starting of the motor without tripping.

Figure 6.13 Pump controllers are the control panels that turn fire pumps on and off and control their operation.

Figure 6.14 The operational components of pump controllers can range from simple to extremely complex.

2. **Isolation switch** — Located between the power supply and the circuit breaker. In some controllers, the isolation switch and the circuit breaker are interlocked so that the isolation switch cannot be operated with the circuit breaker closed. In other cases, the operating handle of the isolation switch is equipped with a spring latch that requires the use of both hands to operate. The isolation switch lever is always located outside the control panel.

3. **Pilot lamp** — Indicates when power is available to the pump control panel. Every controller has a pilot lamp, and it may be found in different locations.

4. **Manual start and stop buttons** — Starts and stops the pump manually. These are standard controller components and are located on the outside of the controller enclosures.

5. **Emergency start lever** — Can be latched in the operating position and provides for continuous nonautomatic operation. This operation is independent of the timer or automatic starting mechanisms.

6. **Running period timer** — Shuts off the motor after the situation has returned to normal. The fire pump operator sets the timer for at least 10 minutes for the monthly tests as required by NFPA® 20. Therefore, if a pump is automatically controlled, it will operate for at least 10 minutes after being turned on unless manually shut down.

7. **Pressure switch** — Turns on a fire pump automatically and is the most commonly used method. It is set to close the circuitry and turn on the fire pump when water pressure on the system drops below the switch setting. Such a pressure drop is usually caused by a sprinkler opening or by hoselines being operated somewhere in the system. The pressure switch is adjustable with high and low settings.

NOTE: Consistent pressure fluctuations may require the use of a pressure-maintenance pump, sometimes referred to as a *jockey pump*, to prevent false pump starts.

If the pump installation is the only source of water supply pressure for a sprinkler system or standpipe system, the automatic start controller must be wired for manual shutdown to ensure that the pump will be turned off only after a fire emergency is over. If the pump room is not constantly attended, transmit audible or visible alarms to an attended location to signal that the pump is running or to signal that the power supply to the pump has been interrupted.

Some pumps may not be equipped with automatic controllers. If manual controllers are used, there will be both a manually operated electric switch to turn the pump on and off, in addition to a mechanical control consisting of a lever or handle that can be latched in the ON position.

Diesel Engine Controllers

The controllers for diesel engines are not interchangeable with the controllers for electric motors **(Figure 6.16)**. Electric motor controllers are designed to start the pumps by closing high voltage circuits. For diesel controllers, the main function is to close the circuit between the storage batteries and the engine starter motor.

Two important features to note are the alarm and signal devices located on the controller itself. If the controller is automatic, a pilot light will indicate when the controller is in the AUTOMATIC position. Separate lights and a common audible alarm are also required to indicate special occurrences such as the following:

- Low engine oil pressure
- High engine coolant temperature
- Failure of the engine to start automatically
- Engine shutdown due to overspeed
- Battery failure
- Low pressure in the storage tanks when air or hydraulic starting is used

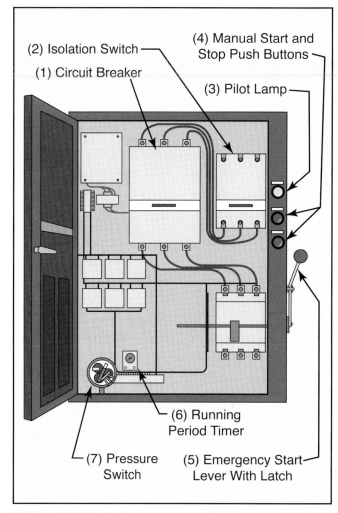

Figure 6.15 Electric fire pump controllers all have the same basic features.

Figure 6.16 While they may appear similar, controllers for diesel engines are not interchangeable with controllers for electric drivers.

NOTE: The controller also requires separate lights for battery-charger failure, but this does not require an audible alarm.

If the pump room is not constantly attended, alarm signals should be transmitted to a constantly attended location. The alarm will indicate when the engine has started, when the controller has been turned off or turned to manual operation, and when trouble exists with the engine or controller.

The controller may also be equipped with a pressure-recording device. The recorder is usually required to run continuously for at least seven days without resetting or rewinding. The recorder's chart drive should be spring-wound, alternating current (AC) electric powered with spring-driven backup, or air-powered.

The provision for automatic starting of the engine due to water pressure is similar to the provision for the electric motor controller. If the pump installation is the sole supply to standpipes or sprinkler systems, the automatic start controller should be wired for manual shutdown. In addition, the automatic controller usually must be arranged to automatically start the engine every week to ensure reliability in engine starting. If the controller is designed to automatically shut down the engine when system conditions return to normal, it should provide running time of at least 30 minutes before shutting the engine off. However, if the overspeed governor operates, the controller will shut off the engine without a time delay.

Pump Components and Accessories

While each pump installation may be different, some basic components are common to all installations. These fit together to make a complete fire pump installation. Pump components and accessories include:

- Piping and fittings
- Relief valves
- Test equipment
- Pressure maintenance pumps
- Gauges

Piping and Fittings

All underground piping that supplies a fire pump must be installed and tested in accordance with NFPA® 24, *Standard for the Installation of Private Fire Service Mains and Their Appurtenances.* These pipes may be constructed of steel, plastic, asbestos cement, or ductile iron.

The aboveground pipe must be constructed of steel **(Figure 6.17)**. The steel pipe can be joined together by screwed or flanged fittings or with grooved fittings. The pipe can also be welded together. Suction piping must also be installed and tested in accordance with NFPA® 24. A control valve of the outside stem and yoke (OS&Y) type must be located in the suction line for control of the water supply.

The proper sizing of both suction and discharge piping is important. If piping size is too small, the pump cannot perform at its rated capacity. The required size depends upon the rating of the pump.

Figure 6.17 All aboveground piping must be constructed of steel. *Courtesy of Ron Moore.*

In most cases, the minimum sizes are the same for both the suction side and the discharge side. However, for a 1,000 gpm (4 000 L/min) pump, the minimum for the suction side of the pump is larger than the discharge side. Although this is not a common requirement, in practice it is common to see the suction pipe larger than the discharge pipe. The larger piping reduces friction loss and increases the pressure available at the pump.

When the size of the suction pipe is different from the inlet port of the pump, an eccentric reducer (not a concentric type) will need to be installed in the suction line. The reason that the reducer must be an eccentric type is to eliminate the possibility of air pockets **(Figure 6.18)**. Air entrapped in the water stream can reduce the efficiency of the pump and can even cause damage due to **cavitation**. The eccentric reducer is to be installed with the flat side on the top.

Cavitation — Condition in which vacuum pockets form due to localized regions of low pressure at the vanes in the impeller of a centrifugal pump and cause vibrations, loss of efficiency, and possibly damage to the impeller.

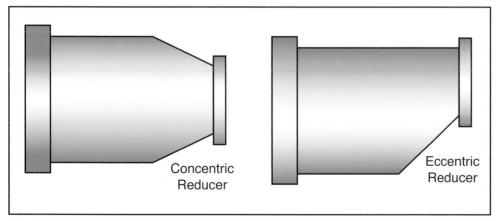

Figure 6.18 An eccentric reducer is used in the suction line to eliminated the possibility of air pockets.

Figure 6.19 Backflow prevention devices are common and are designed to prevent water in the fire suppression system from traveling back into the municipal water supply.

According to NFPA® 20, devices should not be installed that will restrict waterflow on the suction side of a centrifugal pump. Backflow prevention assemblies and check valves may be located on the suction side of the pump if they are located at least ten pipe diameters from the pump intake flange **(Figure 6.19)**. A strainer may be found on the suction side; it is most likely to be used where nonpotable water is supplied. Strainers can also restrict the flow to the pump, especially if they are undersized or dirty.

Relief Valves

Historically, if a pump was driven by a variable-speed driver such as a diesel engine, a **pressure relief valve** was required in the installation **(Figure 6.20)**. More recent versions of NFPA® 20 require these large relief valves only if the pressure at churn is high enough to damage system components. The purpose of this relief valve is to prevent pressures from reaching levels that can damage system piping or fittings. It is possible for an internal combustion engine to lose adjustment and develop excessive speeds. Pump pressure is related to the square of the pump speed. For example, doubling the revolutions per minute increases the pressure developed fourfold. Therefore, the relief valve is provided to open and discharge water to a drain if the pressure becomes excessive. The relief valve should be located between the pump and the discharge check valve. The valve should discharge into an open drain and in a manner so that the waterflow can be visually detected.

The size of the relief valve and its discharge line depends upon the rating of the pump. If the pump was chosen with a proper pressure rating, the relief valve should be set to open a little above the normal discharge pressure when the pump is operating at churn.

A small relief valve, called a **circulation relief valve**, must be provided on electrically driven pump installations and on diesel engine driven pump installations that do not pull cooling water from the pump discharge. It is designed to open and provide enough waterflow into and out of the pump to prevent the pump from overheating when it is operating at churn against a closed system.

Pressure Relief Valve — Pressure control device designed to eliminate hazardous conditions resulting from excessive pressures by allowing this pressure to bypass to the intake side of the pump.

Circulation Relief Valve — Small relief valve that opens and provides enough waterflow into and out of the pump to prevent the pump from overheating when it is operating at churn against a closed system.

Figure 6.20 Pressure relief valves are included in the installation to prevent pressures from reaching levels that would damage piping or fittings.

When the pump is operating at churn, the circulation relief valve should be set to flow a full stream. It should then close off when water begins to flow in the system. The discharge piping should be set to discharge into a drain where the flow can be inspected.

Test Equipment

Every fire pump system is required to have the necessary components for testing the installation. The most prominent of these components is the test manifold **(Figure 6.21, p. 186)**. The test piping should be connected to the pump discharge line between the check valve and the indicating control valve. There should also be an indicating control valve in the test piping. This piping should terminate in a hose valve header located outside the building. The hose valve header should be equipped with the proper hose connections and a shutoff valve for each connection. This manifold and test header will permit water to flow from the pump installation through hoselines and nozzles for test purposes. The flow from the nozzles is measured using pitot tubes.

The required size of the hose header supply pipe and the number of hose valves required depends upon the rating of the pump. It is often possible to estimate the rating of the pump by counting the hose connections. There is usually one 2½ inch (65 mm) hose connection for each 250 gpm (1 000 L/min) of pump rating. For example, a 500 gpm (2 000 L/min) pump usually has two hose connections and a 750 gpm (3 000 L/min) pump usually has three hose connections. This rule of thumb becomes less reliable with 1,250 gpm (5 000 L/min) and 2,000 gpm (8 000 L/min) pumps **(See Table 6.1. p. 186)**.

More modern pump installations do not have the test headers and hose valves. Instead, these installations are equipped with a metering device that can be used to measure the gpm (L/min) delivered by the pump **(Figure 6.22, p. 186)**. This is acceptable, but the meter line should discharge to the outside or back to the water supply source, not directly back to the suction side of the pump. Circulating back to the suction side of the pump will not enable the condition of the suction supply to be evaluated.

Figure 6.21 A typical test manifold.

Table 6.1
Required Number of Test Valves

Pump Rating		Number of Hose Valves	Test Pipe Size	
(GPM)	(L/min)		(Inches)	(mm)
500	2 000	2	4	100
750	3 000	3	6	150
1,000	4 000	4	6	150
1,250	5 000	6	8	200
1,500	6 000	6	8	200
2,000	8 000	6	8	200

Figure 6.22 Metering devices measure the gpm (l/min) delivered by the pump.

Pressure Maintenance Pumps

Sometimes there is enough leakage in the fire protection system or enough fluctuation in the pressure of the water supply to the pump to cause the automatic controller to turn the pump on periodically in nonemergency situations. This can be a nuisance if it happens frequently, particularly in installations that require manual shutdown after automatic operation. A **pressure maintenance pump**, sometimes referred to as a jockey pump, is used to prevent these false starts **(Figure 6.23)**.

Pressure Maintenance Pump — A pump used to maintain pressure on a fire protection system in order to prevent false starts at the fire pump.

Figure 6.23 Pressure maintenance pumps, also called *jockey pumps*, are used to maintain constant pressures without having to activate the main pump. *Courtesy of Ron Moore.*

A pressure maintenance pump is a small-capacity, high-pressure pump used to maintain constant pressures on the fire protection system. This pump takes suction from the fire pump suction line and discharges into the fire pump discharge line on the system side or downstream side of the indicating control valve. The pressure maintenance pump should have adequate capacity to keep up with any leaks in the system. It should be small enough, however, that any demand on the fire protection system, even a single sprinkler opening, will result in the operation of the main fire pump.

The pressure rating of the pressure maintenance pump should be high enough to maintain the desired fire protection system pressure. Set the pressure switch in the fire pump controller to correspond to the system pressure maintained by the pressure maintenance pump. Thus, when the pressure maintenance pump cannot maintain the system pressure due to a demand on the system, the fire pump controller will activate the fire pump. The pressure maintenance pump installation requires the provision of a check valve in the discharge pipe from the pressure maintenance pump as well as indicating control valves. In some cases, pressure relief valves are also required.

Gauges

Most pumps are required to have two gauges: one near the pump intake and one near the pump discharge. Vertical-shaft turbine pumps are only required to have a discharge gauge. These gauges shall be at least 3½ inches (90 mm) inches in diameter and capable of registering pressures of at least 200 psi (1 400 kPa) or twice the rated pressure of the pump, whichever is greater. However, gauges of all sizes and with various pressure ranges are likely to be encountered. The suction gauge must be a compound gauge that registers both positive and negative gauge pressures. Often, an ordinary gauge will register only positive pressures.

Figure 6.24 Liquid-filled gauges can be used in applications where there is significant vibration.

Special attention should be given to pressure gauges. Inaccurate gauges can invalidate test results and give a false indication of how well a pump is performing. During the testing of a pump, the gauge needle may vibrate so widely that an accurate reading is very difficult to obtain. If this is the case, the use of a liquid-filled gauge can eliminate the vibration problem **(Figure 6.24)**.

Pump Location and Component Arrangement

Considerable planning is necessary when selecting the appropriate pump, components, and accessories for a fire protection system. This planning is also necessary when selecting a physical location and arranging these items.

Location and Protection of Fire Pumps

The pump, driver, and controller of a fire pump are required to be protected against possible interruption of service caused by explosion, fire, flood, earthquake, windstorms, freezing, vandalism, or other adverse conditions. At a minimum, outdoor equipment should be shielded by a roof or deck. Fire pump units located outdoors or in buildings other than the building being protected must be located at least 50 feet (15 m) away from the protected building. Indoor fire pump units must be separated from all other areas of the building by fire-rated construction **(Figure 6.25)**.

Heating equipment capable of maintaining a minimum of 40°F (4°C) must be provided in the fire pump room. If a fire protection and detection system is present, the temperature of the pump room must be monitored. The pump room must be provided with proper ventilation. Artificial lighting must be provided in the fire pump location along with emergency lighting. The floor of the fire pump room must be sloped to a floor drain that discharges to a frost-free location. The fire pump room must also be enclosed by fire resistance rated construction (2-hour in accordance with NFPA® 20), or provided in a separate building located at least 50 feet (15 m) away from other buildings.

Component Arrangement

The major components of a pump installation have been identified previously in the chapter. Shown in **Figure 6.26** is a line drawing of a typical fire pump installation using a horizontal split-case centrifugal pump. With this type of pump, the indicating control valves on the supply side of the fire pump and pressure maintenance pump must be OS&Y type. The other control valves shall be of the indicating type.

It may seem as though there are many indicating control valves required in a pump installation. The reason for so many valves is the ability to isolate any component, such as the pressure maintenance pump or the check valve, in the bypass line **(Figure 6.27)**. This enables a component to be repaired or even removed without having to shut off the pump installation. **Figure 6.28, p. 190**, shows a horizontal split-case centrifugal pump installation taking water from a storage tank with a positive head.

Figure 6.25 This fire-pump room is constructed using fire resistant materials.

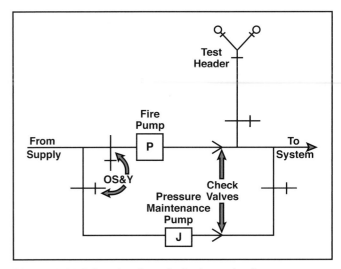

Figure 6.26 A line drawing of a horizontal split-case pump installation.

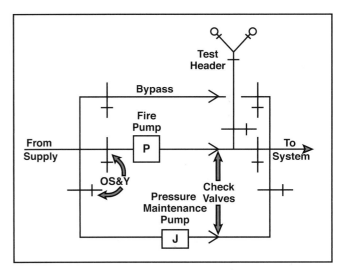

Figure 6.27 Multiple control valves allow for individual components to be isolated in the system should repairs or replacement be necessary.

With a vertical-shaft turbine pump, the components on the discharge side of the pump are essentially the same as they are for the horizontal pump **(Figure 6.29, p. 191)**. There is a relief valve (if needed), the **check valve**, and the test and discharge pipes with their indicating control valves. A vertical-shaft turbine pump often takes suction from a wet pit. This type of installation would ordinarily exist where the water supply source is a pond, lake, or river. An important feature is the double screening required between the pond and the pit that is designed to keep foreign matter out of the pump. The screens should be removable for easy cleaning.

Check Valve — Automatic valve that permits liquid flow in only one direction. For example, the inline valve that prevents water from flowing into a foam concentrate container when the nozzle is turned off or there is a kink in the hoseline.

Testing, Inspection, and Maintenance of Fire Pumps

Fire pumps are extremely important in fire suppression systems because they must be relied upon to provide adequate waterflows when needed. Due to their importance, fire pumps must be properly tested, inspected, and maintained at required intervals in order to ensure proper operation. The following sections address important concepts of fire pump testing, inspection, and maintenance.

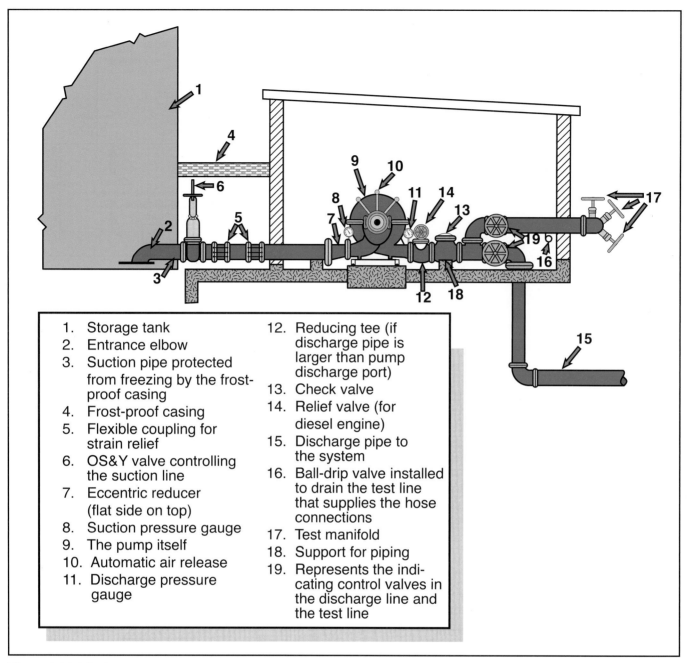

Figure 6.28 A horizontal split-case fire pump installation with water supply under a positive head.

1. Storage tank
2. Entrance elbow
3. Suction pipe protected from freezing by the frost-proof casing
4. Frost-proof casing
5. Flexible coupling for strain relief
6. OS&Y valve controlling the suction line
7. Eccentric reducer (flat side on top)
8. Suction pressure gauge
9. The pump itself
10. Automatic air release
11. Discharge pressure gauge
12. Reducing tee (if discharge pipe is larger than pump discharge port)
13. Check valve
14. Relief valve (for diesel engine)
15. Discharge pipe to the system
16. Ball-drip valve installed to drain the test line that supplies the hose connections
17. Test manifold
18. Support for piping
19. Represents the indicating control valves in the discharge line and the test line

Fire Pump Testing

The primary performance criteria for standard fire pumps are contained in NFPA® 20. The FM Global data sheets also contain information on fire pumps, comparable to NFPA® 20. In order to be considered standard under the provisions of NFPA® 20, a new fire pump must be capable of satisfying the following three test points:

- The pump must not develop more than 140 percent of its rated pressure when operating against a closed system or at churn. (**NOTE:** All horizontal pumps manufactured before 1987 were required not to exceed 120 percent of the rated pressure at churn.)

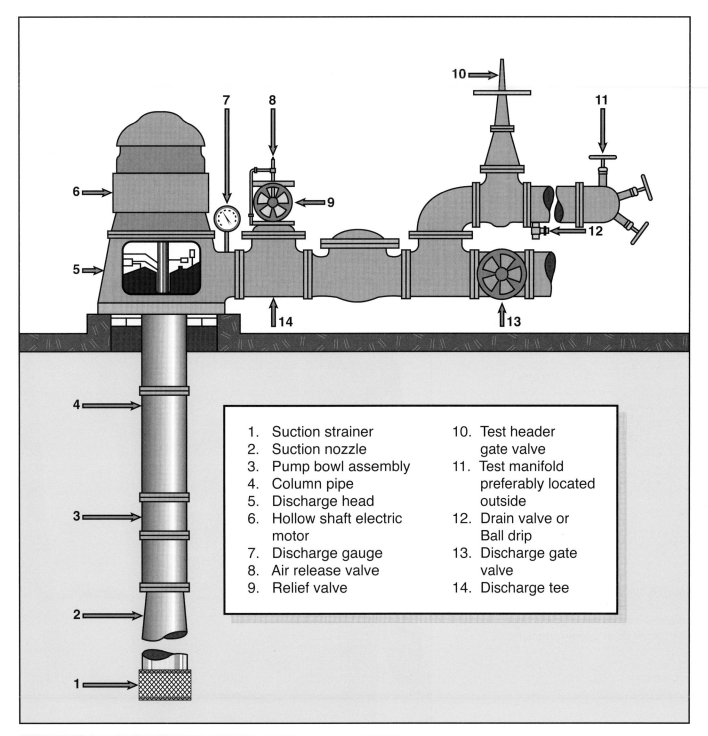

Figure 6.29 A vertical shaft, turbine type fire pump.

1. Suction strainer
2. Suction nozzle
3. Pump bowl assembly
4. Column pipe
5. Discharge head
6. Hollow shaft electric motor
7. Discharge gauge
8. Air release valve
9. Relief valve
10. Test header gate valve
11. Test manifold preferably located outside
12. Drain valve or Ball drip
13. Discharge gate valve
14. Discharge tee

- The pump must operate at minimum of 100 percent of its rated capacity at 100 percent of its rated pressure.

- The pump must operate at 65 percent of its rated pressure while delivering 150 percent of the rated volume in gpm (L/min) or the predetermined maximum volume.

Inspection and Maintenance of Fire Pumps

Fire pumps that fail to operate when needed are likely to result in catastrophic losses. The way to prevent pump failure is to regularly ensure that pump installations are in good operational condition.

Model codes require that pumps be operated at least weekly or monthly, depending upon the type of pump driver. It is not necessary to actually discharge water during the weekly startup. A diesel engine shall be operated for at least 30 minutes, while an electric motor shall be operated for 10 minutes.

Electric motor-driven pumps are often neglected because of their simplicity and low maintenance **(Figure 6.30)**. However, diesel engines require more care and maintenance:

- Follow manufacturer's instructions regarding oil changes. The interval between changes should never exceed one year.

- Check batteries regularly to ensure that they are fully charged.

Figure 6.30 Electric pump drivers are often neglected due to low maintenance requirements, but should be given the same attention as other system components.

- Keep fuel tanks free of water and foreign materials.

- Maintain the temperature in pump rooms that is acceptable to the diesel motor manufacturer.

- Use, if necessary, automatic heaters to keep the engine warm. Keep the engine clean, dry, and well lubricated.

NFPA® 25 requires that an annual flow test of the pump assembly be performed to determine its ability to continue to attain satisfactory performance at shutoff, rated flow, and peak loads. Annual flow tests allow the performance of the pump to be compared year by year.

The manufacturer's recommendations for a preventive maintenance program must be implemented to ensure proper operation of the fire pump. If a maintenance program is not available from the manufacturer, a program that adheres to the requirements as outlined in NFPA® 25 should be adopted. The maintenance program should incorporate a sequence of weekly, monthly, quarterly, and annual tests, along with recommended maintenance that will ensure the continued satisfactory performance of the fire pump assemblies.

Although some types of bearings do not require grease, special attention must be given to clean and lubricate all bearings. The appropriate quantity and the correct type of lubricant must be applied.

Some older pumps may have packing that needs to be checked and adjusted monthly. Others may be equipped with mechanical seals rather than fiber packing. If this is the case and the mechanical seals are in good condition, no leakage should be visible. In addition to the packing and seals, it is a good practice to recheck the pump alignment regularly. Misalignment of the pump drive couplings can cause excessive vibration and damage to the pump.

During weekly inspections, it is also important to make sure that the pump room is kept clean, dry, and free of combustible materials. The weekly inspection is a good time to ensure that all control valves that are supposed to be open are open and are supervised in the OPEN position by padlock and chain, electronic equipment, or tag and seal. This is also the appropriate time to make sure that water tanks and diesel fuel tanks are full.

Personnel will eventually encounter a situation where the pump installation is not operating properly. NFPA® 25 provides a troubleshooting checklist in its Annex C that can help in identifying the causes of pump problems.

Chapter Summary

Fire pump installations are vital parts of a total fire protection system. Their proper selection, installation, and maintenance can be the difference between a quick recovery after a fire and a catastrophe that could destroy life and property.

Fire pumps are used to boost pressure and supply water to fire suppression systems where existing water supplies alone are insufficient. Fire pumps must be able to provide dependable service. In order to do this, they must be properly installed, tested, and maintained.

Review Questions

1. What three actions do fire pumps provide to increase the integrity of a fire suppression system?

2. List the common types of fire pumps.

3. Compare and contrast diesel engine drivers and electric motor drivers.

4. What are the main parts of the pump controller for electric drivers?

5. What is the main function of diesel engine controllers?

6. Name common fire pump components and accessories.

7. What are the requirements for the fire pump room?

8. What are the three test points that a new fire pump must meet?

9. What fire pump maintenance requirements are listed in model codes?

Non-Water-Based Fire Suppression Systems

Chapter Contents

Key Terms

FESHE Outcomes

Fire and Emergency Services Higher Education (FESHE) Outcomes: Fire Protection Systems

1. Explain the benefits of fire protection systems in various types of structures.
4. Identify the different types and components of sprinkler, standpipe, and foam systems.
6. Identify the different types of non-water-based fire suppression systems.
10. Discuss the appropriate application of fire protection systems.

Non-Water-Based Fire Suppression Systems

Learning Objectives

After reading this chapter, students will be able to:

1. Explain the special suppression system classifications.

2. Describe wet chemical fire suppression systems.

3. Explain dry chemical fire suppression systems.

4. Describe clean agent fire suppression systems.

5. Explain carbon dioxide fire suppression systems.

Chapter 7
Non-Water-Based Fire Suppression Systems

Case History

Investigators searching for the cause of a fire that damaged a restaurant in a bed and breakfast facility believe an aerosol can of bug repellent fell into an operating deep-fat fryer and exploded, spewing hot oil around the first-floor kitchen. The resulting fire spread into wall voids.

The two-story, wood-frame building had an asphalt-shingled roof and was protected by a monitored fire alarm system that alerted the occupants. The deep fat fryer was protected by a dry-chemical hood suppression system, but it failed to operate because its cylinder had no pressure. There were no sprinklers. The restaurant's lunch crowd had dispersed 20 minutes before the fire started.

NFPA Journal, May/June 2006, p.30.

Special suppression systems are used in locations where water-based fire suppression systems may not be appropriate for the hazard to be protected. Suppression systems can also be an alternative where preventing water damage is critical in protecting mission essential operations, sensitive electronic equipment, and rare books. These locations might contain food-preparation equipment, combustible metals, or highly sensitive computer or electronic equipment, or water-reactive materials. In these unique locations, water might contribute to the fire, cause additional damage to the building or its contents, or may not effectively control the fire.

Unlike water-based systems, which may have an almost unlimited supply of water, special suppression systems have a limited amount of extinguishing agent. Automatic sprinkler systems are typically designed to control the fire, whereas special agent fire suppression systems are normally intended to extinguish the fire. These systems are also typically designed for a specific hazard or location and a limited area of coverage.

This chapter describes the various classifications and types of special suppression systems, including wet chemical, dry chemical, clean agents, and carbon dioxide (CO_2). In addition, the chapter explains the operations, components, agents, and inspection and testing requirements of these systems.

Special Suppression System Classifications and Types

Suppression systems are classified based upon the class of fire or fires they can effectively extinguish. Fires have been broadly grouped into five classifications according to the burning characteristics of various combustible materials. These classifications include the following:

- **Class A fire** — Involves ordinary combustibles, such as wood, cloth, paper, rubber, and many plastics **(Figure 7.1)**. These fires can be extinguished by cooling, smothering, insulating, or inhibiting the chemical chain reaction.

- **Class B fire** — Involves flammable or combustible liquids and gases, including greases and similar fuels, which can be extinguished by oxygen exclusion, smothering, insulating, and inhibiting the chemical chain reaction **(Figure 7.2)**.

- **Class C fire** — Involves energized electrical equipment, which requires the use of a nonconductive agent for protection of the operator. If the electrical power is eliminated, these fires become Class A or Class B and may be extinguished appropriately.

- **Class D fire** — Involves combustible metals, such as magnesium, potassium, sodium, titanium, and zirconium, which require the use of an agent that absorbs heat and does not react with the burning metal.

- **Class K fire** — Involves cooking oils and fats. Class K-rated agents work by forming a barrier over the product, thus smothering and cooling the fire.

Labels are affixed to the extinguishing system storage tank to indicate the class of fire for which the system is approved.

As previously mentioned, special suppression systems are generally designed with a specific hazard or application in mind. In addition, these systems typically do not serve an entire occupancy, but instead they are designed to protect a specific area. Common types of special suppression systems include the following:

- Wet chemical suppression systems
- Dry chemical suppression systems
- Clean agent suppression systems
- Carbon dioxide suppression systems

Figure 7.1 Class A fires involve ordinary combustibles such as these wooden pallets.

Figure 7.2 Class B fires involve flammable and combustible liquids and gases.

Wet Chemical Fire Suppression Systems

A **wet chemical fire suppression system** is best suited for application in commercial cooking hoods, plenums, ducts, and associated cooking appliances **(Figure 7.3)**. These systems are used in situations where rapid fire knockdown is required. The wet chemical system is most effective when used on fires in deep fat fryers. The nature of the chemical is such that it reacts with animal or vegetable oil and forms soapy foam through a process called *saponification*. This foam forms a blanket that helps to prevent reignition by separating the fuel from oxygen. The wet chemical system should provide a foam blanket that remains intact for at least 20 minutes and not splatter the grease or oil when discharged. Wet chemical agents extinguish grease and oil fires by fuel removal, cooling, smothering, and flame inhibition. All wet chemical systems should meet the requirements set forth in NFPA® 17A, *Standard for Wet Chemical Extinguishing Systems.*

Wet Chemical Fire Suppression System —Suppression system that uses a wet chemical solution as the primary extinguishing agent; usually installed in range hoods and associated ducting where grease may accumulate.

Saponification — A phenomenon that occurs when mixtures of alkaline-based chemicals and certain cooking oils come into contact, resulting in the formation of a soapy film.

Figure 7.3 Wet chemical systems are typically found in commercial kitchen hoods. *Courtesy of Ron Moore.*

Wet Chemical Agents

Wet chemical fire suppression agents are typically composed of water and either potassium carbonate, potassium citrate, or potassium acetate. The agent is delivered to the hazard area in the form of a spray. These are effective suppression agents for fires involving flammable/combustible liquids such as grease or oil, or ordinary combustibles such as paper and wood.

More recently introduced, alkaline mixtures are useful for attacking Class K fires because of their ability to generate a soapy, foam-like film. The agents used in this system are the same as those described for portable fire extinguishers in Chapter 9, Portable Fire Extinguishers.

Wet chemical systems can be messy, particularly when food greases are involved. The spray from a wet chemical system can also migrate to surrounding surfaces, causing corrosion of electrical wires or damage to hot cooking equipment. For this reason, it is important to ensure a prompt cleanup after activation of a wet chemical system. This type of system is not recommended for electrical fires because the spray mist may act as a conductor.

System Components

Wet chemical systems typically have the same components. These components are as follows:

- **Storage tank or tanks (Figure 7.4)** — The storage container may contain both the agent and the pressurized expellant gas, or the agent and the gas may be stored separately. A pressure gauge attached to the container is an indication of a stored-pressure container. The expellant gas used is typically either nitrogen or carbon dioxide. System tanks must be located as close to the discharge point as possible, but they also must be in a temperature-controlled area.

- **Piping** — A system of fixed piping is used to carry the gas and agent. The piping is specially designed to account for the unique flow characteristics of the agent. The proper pipe size, number of bends and fittings, and pressure drop (friction loss) are all taken into account when calculating piping requirements.

- **Nozzles (Figure 7.5)** — The wet chemical agent is dispersed on the hazard through nozzles. Nozzles are attached to the piping.

- **Actuating mechanism** — Wet chemical is released into the piping system in response to activation devices. These devices are usually designed to activate when fusible links melt due to heat. The fusible links trigger a mechanical or electrical release that in turn starts the flow of expellant gas and agent **(Figure 7.6)**. Another method of automatic activation includes pressurized pneumatic tubing. Fixed systems should be capable of manual activation and must be equipped with automatic fuel or power shutoffs. The shutoff device must be restored manually.

NOTE: Detailed maintenance procedures and restoration of these systems should be left to trained personnel.

Figure 7.4 Storage tanks for both the extinguishing agent and the expellant gas are located adjacent to the system.

Figure 7.5 Nozzles in a typical wet chemical system.

Figure 7.6 Fusible links in the system are activated by heat and trigger a release that starts the flow of extinguishing agent.

Inspection and Testing Procedures

A record should be kept of each inspection and any problems noted. The system service provider should immediately correct any problems. A competent and trained individual should inspect most systems on a semiannual basis as required by model codes. A trained individual should check that all components work properly, piping is unobstructed and that there is no evidence of corrosion, structural damage, or repairs. The liquid levels in nonpressurized containers should also be checked. Trained personnel should be able to inspect these systems for:

- Mechanical damage
- Aim of nozzles
- Changes in hazards
- Proper pressures in stored-pressure containers
- Maintenance tags for scheduled service

Dry Chemical Fire Suppression Systems

Dry chemical fire suppression systems are used wherever rapid fire suppression is required and where reignition of the burning material is unlikely. These systems are most commonly used to protect the following areas:

• Flammable and combustible liquid storage rooms

• Dip tanks

• Spray paint booths **(Figure 7.7)**

• Old commercial cooking areas or kitchens

• Exhaust duct systems

• Heavy equipment, such as earthmovers and generators

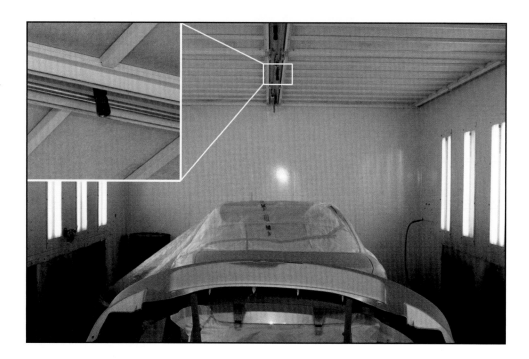

Figure 7.7 Dry chemical systems are commonly found in paint spray booths.

Dry Chemical Systems for Protection of Kitchen Equipment

Most existing dry chemical systems installed to protect kitchen equipment do not meet listing requirements (ANSI/UL 300, *Fire Testing of Fire Extinguishing Systems for Protection of Commercial Cooking Equipment*) for dry chemical fire suppression systems for commercial kitchen hood, duct, and cooking appliances. Failure to meet the listing requirements required by NFPA® 17 makes repair, annual certification, and testing difficult. This failure results in requiring the dry chemical systems protecting kitchen equipment to be removed from service and replaced with a compliant system.

All dry chemical systems should meet the requirements set forth in NFPA® 17. The components of dry chemical systems are virtually the same as those for wet chemical systems. Two methods used for the application of dry chemical extinguishing agents are fixed systems and handheld hoselines; however, this manual will only address fixed systems.

Fixed Dry Chemical Systems

Fixed dry chemical systems consist of the agent storage tanks, expellant storage tanks, a heat-detection and activation system, piping, and nozzles. Two main types of fixed systems are:

- **Local application** — Discharges agent onto a specific surface, such as the cooking area in a restaurant kitchen. The most common type of fixed system that is no longer code compliant or listed for kitchen use, but they may still be encountered protecting small industrial hazards such as a dip tank.

- **Total flooding** — Introduces a thick cloud of agent into a closed area, such as a spray paint booth.

Dry Chemical Agents

Dry chemical suppression systems discharge a cloud of chemical that leaves a residue. This residue creates cleanup problems after system operation. A dry chemical system is not recommended for use in areas that contain sensitive electronic equipment. The chemical residue has insulating characteristics that hinder the operation of the equipment unless extensive cleanup is performed. The agent also becomes corrosive when exposed to moisture. Two dry chemical extinguishing agents that are used in dry chemical extinguishing systems include:

- **Sodium bicarbonate** — Also known as *ordinary dry chemical*; sodium bicarbonate is effective on Class B and Class C fires. Sodium bicarbonate is twice as effective as an equal amount of carbon dioxide for Class B fires. This agent is also used for Class A fires in textile machinery where the fibers can produce a surface fire. Sodium bicarbonate used in fire extinguishing systems is chemically treated to be water repellant and free-flowing.

- **Monoammonium phosphate** — Also known as *multipurpose dry chemical*; monoammonium phosphate is effective on Class A, Class B, and Class C fires. This agent has an action similar to other dry chemicals on flammable liquid fires. Using a combination of extinguishing methods, it quickly extinguishes flaming combustion. When used on Class A materials, monoammonium phosphate melts, forming a solid coating that extinguishes the fire by smothering. Monoammonium phosphate is the most corrosive of dry chemical agents and can have a corrosive effect on unprotected metals. Corrosion can form around extinguisher system nozzles, piping, and agent containers.

Inspection and Testing Procedures

Personnel should keep a record indicating that each inspection was made and immediately correct any problems that were noted during inspection. In most cases, correction of deficiencies requires notification of the fire protection system company responsible for the maintenance of the system. Personnel should conduct inspections as required by NFPA® 17 which has recommendations for monthly and semiannual inspection and testing.

Clean Agent Fire Suppression Systems

Clean agent fire suppression systems are used in areas where water-based suppression systems or wet chemical or dry chemical systems may be undesirable or unsuitable **(Figure 7.8)**. Clean agent suppression systems, acting as a first line of defense against fire, are also used in conjunction with water-based fire suppression systems as a backup. These areas must undergo room-integrity testing to ensure that the agent will be retained within the room to accommodate fire extinguishment. Clean agent systems store the extinguishing agent as a liquid. When the agent is exposed to the atmosphere, it turns to gas. In some cases, the gas extinguishes fire by displacing oxygen. In other cases, the gas disrupts the chemical chain reaction of the fire. Depending on the agent, the area may become untenable for occupants due to potentially lowered oxygen levels. Clean agents are effective on Class A, Class B, and Class C fires and will not conduct electricity.

Some typical applications for clean agent fire suppression systems include the following:

- Computer rooms
- Telecommunications facilities
- Clean (manufacturing) rooms
- Data storage areas
- Irreplaceable document and art storage rooms
- Laboratories
- Art galleries and museums
- Boats and vehicles

In order for clean agent systems to work effectively, the agent must be maintained at a certain concentration within the room for a given period of time. If the room is not sealed, the agent will escape the room and the required concentration will not be maintained as required for fire extinguishment. Open doors, open ductwork, and even open conduit penetrations or holes in the walls could be detrimental to proper operation of the system. In those areas where these systems are used, there are automatic door closers, door sweeps, and **predischarge warning devices (Figure 7.9)**. These warning devices allow the occupants in those areas to evacuate the space or room before the system discharges. Placarding or identification signage is important in areas where these systems are installed. In addition, any heating, ventilating, and air-conditioning (HVAC) systems must be controlled to ensure room integrity.

Clean Agents

Clean agents are in a general category of fire extinguishing agents that leave no residue. These agents extinguish by smothering the burning material, excluding oxygen, and interrupting the chemical chain reaction of the fire. The U.S. Environmental Protection Agency (EPA) has approved modern clean agents as nonharmful to the atmosphere.

Halogenated fire extinguishing agents were one of the first groups of clean agents developed. Halogenated agents are principally effective on Class B and Class C fires. The word **halon** has been commonly used to describe this group of agents.

Figure 7.8 This small clean-agent system is incorporated to protect a small records room.

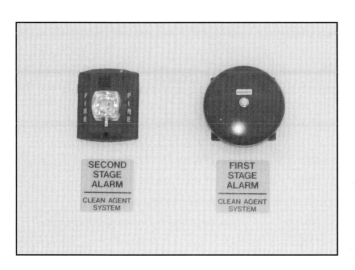

Figure 7.9 Predischarge warning devices for clean-agent release may include multiple stages. *Courtesy of Texas Fire Prevention Specialists*.

While halon agents are effective, they have been proven to be harmful to humans and the earth's ozone layer, so restrictions have been placed on their production. Although the Montreal Protocol of 1987 provided for a phase out of halon agents and forbade manufacture of new halon agents after January 1, 1994, limited production continues because of some exceptions to the phase-out plan.

Two types of halons are still in use:

- Halon 1211 is most commonly found in portable extinguishers.

- Halon 1301 is used in some portable fire extinguishers but is more commonly found in fixed fire extinguishing systems for total flooding applications.

There are products that extinguish fires in the same manner as halogenated extinguishing agents and are less damaging to the atmosphere. These products are known as *halon-replacement agents*.

Common halon-replacement agents include the following:

- Halotron® — Clean agent that when discharged is a rapidly evaporating liquid. Halotron® leaves no residue and meets (EPA) minimum standards for discharge into the atmosphere. The agent does not conduct electricity back to the operator, making it suitable for Class C fires. Halotron® has a limited Class A rating.

- FM-200® — Hydrofluorocarbon that is considered to be an acceptable alternative to Halon 1301 because it leaves no residue and is not harmful to humans and the environment. This agent does require significantly more agent for effective extinguishment than Halon 1301. There are several derivatives of FM-200® that have slightly different compositions.

- ECARO-25 — Hydrofluorocarbon is a safe, clean, and electrically nonconductive agent suited for use at low temperatures. This agent presents similar flow characteristics to Halon 1301 and is often used as a direct replacement for the ozone-depleting substance.

- Inergen® — Blend of three naturally occurring gases: nitrogen, argon, and carbon dioxide. It is stored in cylinders near the facility under protection. Inergen® is environmentally safe and does not contain a chemical composition like many other proposed halon alternatives.

- Novec™ 1230 — Halon replacement that is stored in a liquid state but converts to a gas when discharged through a nozzle. Novec™ 1230 is environmentally sustainable and suited for sensitive areas, such as computer rooms and museums.

   NOTE: Refer to **Table 7.1**, Names of Chemical Compounds.

Table 7.1 Names of Chemical Compounds	
ECARO-25	Pentafluoroethane
FM-200®	1,1,1,2,3,3,3-Heptafluoropropane
Halon 1211	Bromochlorodifluoromethane
Halon 1301	Bromotrifluormethane
Halotron®	hydrochlorofluorocarbon
Novec™ 1230	dodecafluoro-2-methylpentan-3-one

Clean Agent System Components

Clean agent suppression systems may be fixed systems that are designed for local application or total flooding agent distribution. Components for these systems are similar to other systems previously explained in this chapter. Clean agent system components include agent storage containers, piping, and discharge nozzles. The system is used in conjunction with an approved fire alarm releasing system that has initiating devices and notification appliances.

Inspection and Testing Procedures

Inspection and testing procedures for clean agent fire extinguishing systems can be found in NFPA® 2001, *Standard on Clean Agent Fire Extinguishing Systems*. Qualified personnel are required to perform an annual inspection for all system components. Personnel must semiannually check the quantity and pressure of the clean agent. Records on all tests and inspections must be maintained and made available for the (AHJ) inspectors to review.

Personnel must inspect the protected enclosure annually, in addition to system inspections and tests. The inspector must determine that the integrity of the space has not been compromised by penetrations that would permit the agent to escape during a discharge.

Carbon Dioxide Fire Suppression Systems

Carbon dioxide (CO_2) is a type of clean agent that has been proven effective for extinguishing most combustible materials fires with the exception of some reactive metals, metal hydrides, and other materials that contain available oxygen. The limitations of CO_2 are related to health effects associated with its use as well as restrictions imposed by the combustible material itself.

As delivered, CO_2 is extremely cold [approximately -110°F (-79°C)] and can freeze exposed skin. The agent, however, has a limited cooling effect on a fire. The primary mechanism of extinguishment of CO_2 is accomplished through oxygen removal or smothering. The cooling effects of its application (although small) are realized when the agent is applied directly to the burning material.

CO_2 fire suppression systems closely resemble clean agent systems. All systems must adhere to the requirements as described in NFPA® 12, *Standard on Carbon Dioxide Extinguishing Systems*. These systems have been used to extinguish fires involving the following materials or equipment:

- Flammable and combustible liquids
- Electrical equipment and energized equipment
- Flammable gases
- Other combustibles, including cellulose materials
- Engine test cells **(Figure 7.10, p. 210)**

Although CO_2 systems were almost phased out by the growth of halon and subsequent clean agent systems in the 1970s and 1980s, there has been a resurgence of these systems as the need to replace halon-based systems has increased. A CO_2 fire suppression system can be used to protect a wide variety of

Figure 7.10 CO_2 fire suppression systems are used to extinguish fires involving the use of engine test cells.

hazards through total flooding or local fire protection applications. Handheld hoseline and standpipe systems are also used, although they are uncommon. Some types of hazards that CO_2 can protect include:

- Automobile manufacturing facilities
- Industrial plants
- Refineries/chemical plants
- Paint and coating operations
- Food/agricultural processing plants
- Pharmaceutical manufacturing facilities
- Printing presses
- Gas turbines

Personnel safety is the most serious problem involving CO_2 systems, especially total flooding systems. The elimination of oxygen from a fire also eliminates breathable oxygen from the atmosphere, creating an asphyxiation hazard for personnel in the area. Total flooding systems must be provided with pre-discharge alarms as well as discharge alarms. A pre-discharge alarm notifies those individuals present that the system is about to activate, indicating the need to immediately evacuate the area. Total flooding systems must also be located within an enclosure. This enclosure must either be sealed to prevent the leakage of CO_2 or additional CO_2 capacity must be provided to account for any leakage. NFPA® 12 contains retroactive requirements for the installation of an odorizer and pneumatic alarm on existing systems.

Carbon Dioxide System Components

The components of the CO_2 system are similar to other special suppression systems. These components include agent storage containers, piping, and nozzles. The system is used in conjunction with an approved fire alarm releasing system that has initiating devices and notification appliances.

The three means of actuation for CO_2 systems are as follows:

- **Automatic operation** — Triggered by a product-of-combustion detector, such as smoke detectors, fixed-temperature, or rate-of-rise detection equipment.

- **Normal manual operation** — Activated by a person manually operating a control device and putting the system through its complete cycle of operation, including predischarge alarms.

- **Emergency manual operation** — Used only when the other two actuation modes fail, causing the system to discharge immediately and without any advance warning to individuals in the area.

CO_2 systems are available in high-pressure and low-pressure systems. In a high-pressure system, the CO_2 is stored in standard U.S. Department of Transportation (DOT) approved cylinders at a pressure of about 850 psi (5 860 kPa) **(Figure 7.11)**. A low-pressure system is designed to protect much larger hazards. The liquefied CO_2 in these systems is stored in large, refrigerated tanks at 300 psi (2 068 kPa) at a temperature of 0°F (-18°C).

In both systems, the containers are connected to the discharge nozzles through a system of fixed piping. Nozzles for total flooding systems may be the high- or low-velocity types. High-velocity nozzles promote better disbursement of the agent throughout the entire area. Local application nozzles are typically the low-velocity type, which reduces the possibility of splashing the burning product when it comes in contact with the agent.

Figure 7.11 Carbon dioxide systems are stored at either high or low pressure. System pressure can be determined by reading the pressure gauges on the system.

Inspection and Testing

Due to the complexity of CO_2 systems, fire inspectors or personnel should limit inspection items to the following:

- Physical damage of the components
- Excessive corrosion

- Change in hazard
- Enclosure integrity
- Up-to-date test, maintenance, and inspection records

Chapter Summary

Special fire suppression systems are needed when the application of water is not the best way to extinguish a fire due to the nature of the hazard itself or because large amounts of water will ruin specialized equipment. In these situations, better choices may be the specialized and controlled application of wet or dry chemicals, CO_2, or clean agents. Because these systems are often custom-designed and contain only a specific amount of extinguishing agent, they must be installed carefully and inspected regularly to ensure successful operation. Many of the chemicals and CO_2 used in specialized systems are dangerous. Therefore, people who may be affected must be made aware of the potential hazards. Individuals must know how to exit an area promptly and activate the fire suppression system from a position of safety.

Review Questions

1. How are special suppression systems classified?

2. A wet chemical fire suppression system is best suited for application under what conditions?

3. Name the extinguishing agents used in dry chemical fire suppression systems.

4. List the common halon-replacement agents.

5. What are the limitations of CO_2 fire suppression systems?

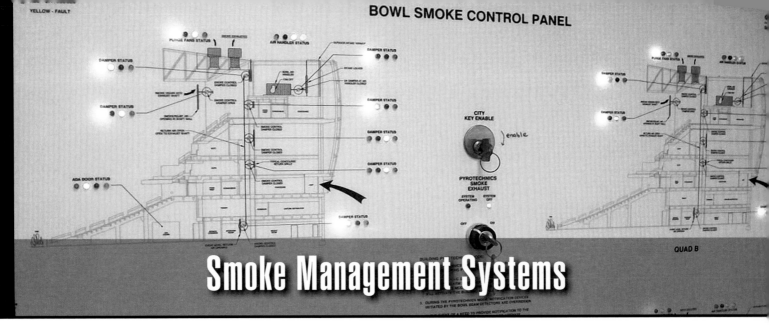

Smoke Management Systems

Chapter Contents

Courtesy of Sprint Center, Kansas City, Missouri.

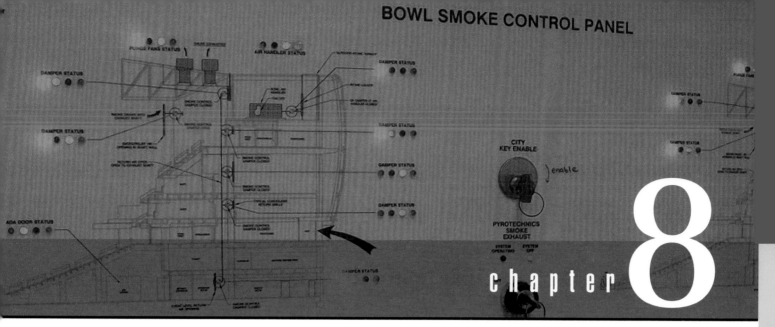

Key Terms

FESHE Outcomes

Fire and Emergency Services Higher Education (FESHE) Outcomes: Fire Protection Systems

1. Explain the benefits of fire protection systems in various types of structures.

9. Describe the hazards of smoke and list the four factors that can influence smoke movement in a building.

10. Discuss the appropriate application of fire protection systems.

Smoke Management Systems

Learning Objectives

After reading this chapter, students will be able to:

1. Describe hazards associated with smoke and other products of combustion.

2. Identify factors that affect smoke generation and spread in buildings.

3. Explain the purpose of a smoke management system.

4. Summarize the history of smoke management systems.

5. Describe the strategies and types of smoke control in buildings.

6. Explain the purpose and functions of a firefighters' smoke control station (FSCS).

7. Summarize types of testing for smoke management systems.

8. Describe smoke management system implications for emergency services personnel.

Chapter 8
Smoke Management Systems

Case History

In July 2015, a fire broke out in a 14-story apartment building in New Jersey. The fire originated in an office on the first floor of the high-rise apartment complex. The fire was contained to the first floor of the building; however, smoke spread throughout the building forcing the evacuation of 350 residents. None of the 350 residents were injured. Due to the age of the building, it was not equipped with a smoke management system or a fire sprinkler system. The movement of smoke in the building was due in large part to the HVAC and utility pipe chases.

Fires can generate enormous amounts of smoke, and it is smoke that kills the majority of fire victims **(Figure 8.1, p. 218)**. In many reported instances, victims were found deceased in areas remote from the fire. In addition, smoke contributes to overall property damage well beyond the origin of the fire. Smoke control/exhaust systems mechanically remove smoke and other products of combustion from the structure.

Contemporary fire protection may include smoke management or smoke control as a part of the overall fire protection and detection systems of the building. This is especially true in occupancies such as high-rise buildings, covered malls, buildings with atriums, and warehouses with high-piled storage. Smoke management is an all-inclusive term that can include compartmentation, pressurization, exhaust, dilution, and buoyancy elements. Smoke management also includes smoke barriers and exhaust fans and vents **(Figure 8.2, p. 218)**. *Smoke control* refers to any effort to change the pressure in spaces adjacent to the fire area to compartmentalize or exhaust smoke from the area of the fire's origin.

The design of **smoke control systems** is a relatively complicated process involving complex mathematical calculations that have been developed from a variety of tests and experiments and computer modeling. This chapter serves as an introduction to the concept of smoke management and control and the variety of strategies that can be used in the accomplishment of this lifesaving technique. **Smoke management systems** are currently addressed in NFPA® 92, *Standard for Smoke Control Systems*. Additional information pertaining to smoke management systems can be found in model codes and American Society of Heating, Refrigerating, and Air-Conditioning Engineers (ASHRAE) Standard 52, and the *Handbook of Smoke Control Engineering*.

Smoke Control System — Engineered systems designed to control smoke by the use of mechanical fans to produce airflows and pressure differences across smoke barriers to limit and direct smoke movement.

Smoke Management System — System that limits the exposure of building occupants to smoke. May include a combination of compartmentation, natural or mechanical means to control ventilation and smoke migration from the affected area, as well as a means of removing smoke to the exterior of the building.

Figure 8.1 Smoke generated by fires can often be substantial in quantity and extremely toxic.

Figure 8.2 Smoke vents are incorporated into a smoke management system in order to efficiently remove harmful smoke from the structure. *Courtesy of Texas Fire Prevention Specialists.*

Smoke and Other Products of Combustion

The dangers of smoke and its movement during a fire are related to the products of combustion and the hazards that are created by smoke. Products of combustion can be described very simply as heat, smoke, and sometimes light. However, this description is very deceptive. As fuel burns, its chemical composition is altered, which results in the production of new substances and the release of energy. Structure fires typically involve multiple types of combustible materials and a limited amount of oxygen, which results in incomplete combustion. These factors culminate in the form of complex chemical reactions that produce a wide range of products of combustion. Some of these include toxic and flammable gases, vapors, and particulates.

Heat is a product of combustion that contributes to the spread of fire. This occurs through the preheating of adjacent fuels through **pyrolysis** that makes them more susceptible to ignition **(Figure 8.3)**. Heat is also a significant cause of injury. The heat from a fire causes burns, damage to the respiratory system, dehydration, and heat exhaustion.

Smoke is an aerosol comprised of fire gases, aerosols, and solid particulates. Fire gases such as carbon monoxide (CO) are generally colorless, while

Pyrolysis — Thermal or chemical decomposition of fuel (matter) because of heat that generally results in the lowered ignition temperature of the material.

aerosols and particulates in smoke can give it a variety of colors. Most components of smoke are toxic and present a significant threat to human life. The materials that compose smoke vary depending on the fuel, but generally all smoke is toxic.

Irritants in smoke are those substances that cause breathing difficulty and inflammation of the eyes, respiratory tract, and skin. Depending on the fuels involved, smoke will contain a wide range of irritating substances.

The toxic effects of smoke inhalation are not the result of any one gas. Instead, these are interrelated effects of all the toxic products present. Three of the more common products of combustion that can be hazardous to building occupants and firefighters include:

- **Carbon monoxide (CO)** — By-product of the incomplete combustion of organic (carbon-containing) materials. CO is probably the most common product of combustion encountered in structure fires. Exposure to CO is frequently identified as the cause of death for civilian fire fatalities and firefighters who have run out of air in their self-contained breathing apparatus (SCBA).

- **Hydrogen cyanide (HCN)** — Produced in the combustion of materials containing nitrogen and is a significant by-product of the combustion of polyurethane foam (commonly used in furniture and bedding). HCN is also commonly encountered in smoke, although at lower concentrations than CO.

- **Carbon dioxide (CO_2)** — Product of complete combustion of organic materials and is not toxic in the same manner as CO and HCN. CO_2 acts as a simple asphyxiant by displacing oxygen. CO_2 also acts as a respiratory stimulant by increasing the respiratory rate.

Other hazardous components can be found in smoke depending upon the composition of the fuel that is burning. As previously mentioned, toxic smoke is responsible for the majority of fire deaths due to the combination of chemicals contained in the smoke. CO causes asphyxiation because it binds much better than oxygen in the blood. Therefore, oxygen is displaced from the hemoglobin in the blood, reducing the amount of oxygen available for use by the body.

Dense smoke also causes visual impairment and can limit the ability of the building's occupants to evacuate. In addition, smoke presents a more hazardous environment for fire fighting personnel **(Figure 8.4, p. 220)**.

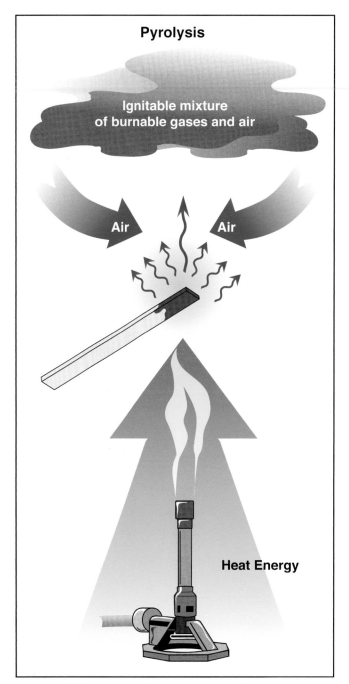

Figure 8.3 Pyrolysis is the preheating of adjacent fuels that makes them more susceptible to ignition.

Figure 8.4 A hazardous environment for firefighters is working in dense smoke.

Smoke Movement in Buildings

During a fire, smoke will move throughout the building while following the laws of physics and the path of least resistance. Although a fire may be confined within a fire-resistive compartment, smoke can spread easily to adjacent areas through openings such as construction cracks, pipe penetrations, ducts, open doors, and vertical shafts. The heating, ventilating, and air-conditioning (HVAC) system in a building can also provide a path for the communication of smoke through a building.

Smoke generation and spread can be affected by several factors. These factors include the following:

- Stack (chimney) effect
- Buoyancy
- Weather
- Mechanical air-handling systems

Some of these factors can cause pressure differences between partitions, compartments, walls, and floors that can result in the spread of smoke. Smoke management systems are designed to take these factors into consideration.

Stack (Chimney) Effect

The **stack effect** is a naturally occurring vertical movement of air within a building. This phenomenon is also referred to as the *chimney effect*. Stack effect occurs due to a difference in the temperature of a building's interior and the exterior **(Figure 8.5).** When the outside air is cooler than the building's interior, air will move upward through the building in large openings, atriums, stairwells, and vertical shafts. During a fire, smoke movement occurs upward from the fire below through open shafts. Smoke then flows out of the shafts and into the upper floors of a building. This is sometimes referred to as a *normal stack effect*.

Stack Effect—
Phenomenon of a strong air draft moving from ground level to the roof level of a building. The air movement is affected by building height, configuration, and temperature differences between inside and outside air.

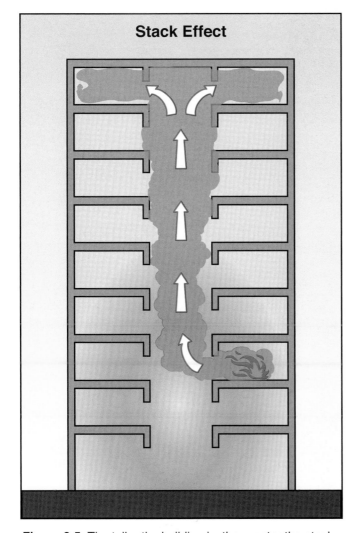

Stack Effect

Figure 8.5 The taller the building is, the greater the stack effect.

A downward flow of air can occur during warmer weather when the air inside is cooler than the air outside. During very hot weather conditions, the stack effect causes air to move vertically downward in buildings and is referred to as *negative* or *reverse stack effect*. Since stack effect is a function of temperature, it is of greatest concern when the temperature differential between the inside and outside of the building is high.

Buoyancy

Smoke **stratification** and movement occurs in part due to the smoke's buoyancy. This buoyancy is due to the

Stratification — Formation of smoke into layers as a result of differences in density with respect to height with low density layers on the top and high density layers on the bottom.

smoke being a higher temperature than the surrounding air. The higher temperature reduces the density of smoke and causes it to expand, thus leading to smoke movement throughout the structure **(Figure 8.6)**.

In an unsprinklered building, the buoyancy of the fire gases can be a significant factor in the movement of smoke in the building. Smoke forms a layer in the upper part of the fire compartment and spreads vertically out through openings to the floors above. In a sprinklered building, the buoyancy factor is typically reduced due to the cooling of the smoke by the water from the sprinkler.

Weather

Weather can have an effect on smoke movement. While the effect of temperature was discussed previously, wind can have a pronounced effect on smoke movement as well **(Figure 8.7)**. The pressure that wind exerts on the walls of a building depends on the building geometry and wind obstructions. It is possible to calculate wind pressures based on the design of the building.

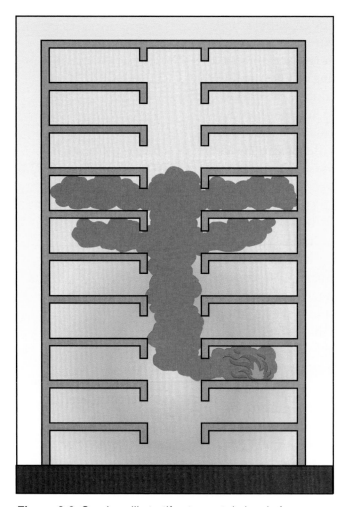

Figure 8.6 Smoke will stratify at a certain level of a structure when its temperature decreases.

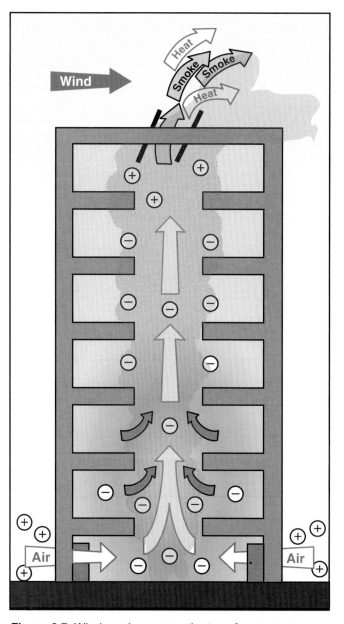

Figure 8.7 Wind moving across the top of a structure can increase stack effect.

Buildings that are exposed to the wind experience a positive wind pressure on the wall facing the wind and a negative wind pressure on the other three sides. The shape of the building and the velocity of the wind both influence wind pressure. Wind acts to promote horizontal, rather than vertical, air movement through the building. This spreads smoke from the windward side to the leeward side of the building. Of course, in a building with tight exterior walls and no windows, the effects of wind on smoke spread are more likely to be minimal. However, when windows are broken or other large openings are created in exterior walls, the wind will have a greater effect. The force of the wind acting upon the upper stories of a high-rise building may be particularly strong and lead not only to smoke being forcefully pushed through the building, but to extreme fire behavior.

Mechanical Air-Handling Systems

HVAC systems can also contribute to smoke movement through a building **(Figure 8.8)**. When a fire starts, an operating HVAC system can transport smoke to every area that the system serves **(Figure 8.9, p. 224)**. As the fire progresses, the smoke can endanger life, damage property, and inhibit fire fighting activities. This is especially true with HVAC systems that are not shut down during a fire or that recirculate air through the building. Even if the system is shut down, this alone will not prevent smoke movement through the shafts and return ducts in the system. HVAC ductwork systems are often equipped with duct smoke detectors that will deactivate fans and shut smoke dampers in an effort to limit smoke spread. However, these ductwork systems may cause another avenue for smoke spread from floor to floor.

Figure 8.8 HVAC systems can contribute to the movement of smoke to other areas of the building.

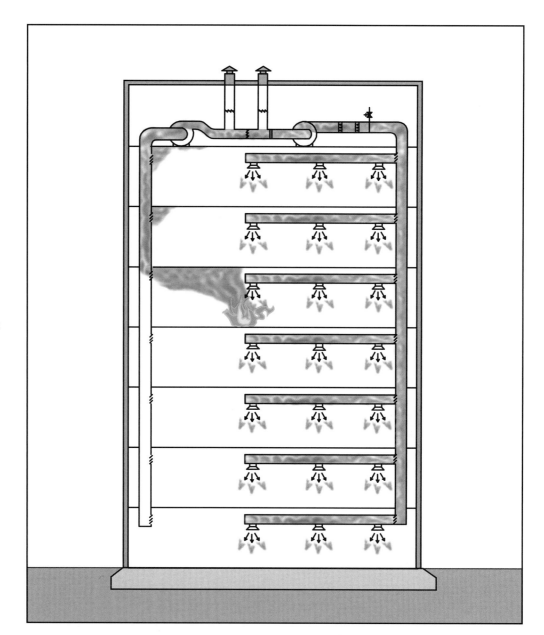

Figure 8.9 The HVAC system can draw the products of combustion into the ducts and transport them throughout the building.

Purpose of a Smoke Management System

A smoke management system reduces occupant deaths and injuries and aids in the safety of firefighters by removing and controlling the spread of smoke. An added benefit of these systems is the reduction of property loss from smoke damage. Most systems are designed for life safety purposes, though some may be more for property protection, especially where high-value contents are at risk. Smoke management systems are designed to provide a safe escape route, a safe refuge area, or both.

Smoke management systems accomplish their objective in one or more of the following ways:

- Maintain a tenable environment in the area of egress during the time required for evacuation

- Control and reduce the migration of smoke from the fire area

- Provide conditions outside the fire zone that will assist emergency response personnel in conducting search-and-rescue operations and in locating and controlling the fire (**Figure 8.10**)

History of Smoke Management Systems

The need for smoke management systems dates back to the late 19th and early 20th centuries when large numbers of individuals were killed in several major theater fires. Those incidents included the Brooklyn Theater fire (283 died), the Vienna Ring Theater fire (449 died), and the Theater Royal fire (186 died).

In 1911, a fire broke out in the Empire Palace Theatre in Edinburgh, Scotland, United Kingdom. The outcome of this fire was different. Smoke vented through the stage roof, which was credited with the prevention of any loss of life. The buoyancy of the hot smoke forced it through the vent openings.

The 1980 MGM Grand Hotel fire in Las Vegas, Nevada, is a more modern-day example of the harm that can result from smoke spread. Most of the deaths from this incident were in the upper floors of the hotel. Smoke from the fire on the first floor spread upward through the HVAC system, the elevator hoistways, and unprotected vertical openings.

In 1973, the Building Department of the City of Atlanta, Georgia, conducted a series of tests of fire protection and detection systems in the Henry Grady Hotel, a 14-story building that was scheduled to be demolished. The purpose of the tests was to evaluate the effectiveness of various smoke-control systems, including stairwell pressurization (both with and without vestibules) and elevator hoistway pressurization. The stairwell systems were intended to provide a smoke-free egress for building occupants, and the elevator system was intended to prevent smoke movement. The Henry Grady Hotel project demonstrated that pressurization could provide a relatively smoke-free environment for egress given the fire scenarios and systems tested.

Additional tests were also performed at the Church Street Office Building in New York City, New York (1973) and a seven-story office building in Hamburg, Germany (1976). In each of these test programs, the smoke management systems were proven to be effective in managing the spread of smoke.

Figure 8.10 Stairwells often provide an area free of smoke for firefighters to stage equipment and prepare to enter the hazardous area.

Smoke Control Strategies

A variety of smoke control methods or strategies exist. Some are installed as dedicated systems, while others can serve different purposes for the building systems during normal building operations. Different strategies for smoke control include:

- Passive systems (including compartmentation)
- Pressurization systems
- Exhaust method
- Opposed airflow method
- Dilution
- Zoned smoke control

Dedicated smoke control systems are those that are intended and specifically listed for smoke control purposes **(Figure 8.11)**. These systems allow for separate air movement and distribution and do not function under normal operating conditions in the building. Upon activation, these systems operate specifically for smoke control.

Advantages of a dedicated smoke control system include:

- Operation and control are generally simpler than other systems.
- Modification of the controls during system maintenance is less likely to occur than with other systems.
- It is less likely to be affected by the modification or failures of other building systems.

While dedicated systems have significant advantages, they may be more costly, require more building space, and have unresolved operational failures since the equipment is not used during normal building operations. Therefore, a strong administrative control and inspection program should be in place.

Nondedicated smoke control systems are those that share components with other systems such as the building's HVAC system **(Figure 8.12)**. Activation causes the system to change its mode of operation for smoke control purposes **(Figure 8.13)**.

Potential advantages of nondedicated smoke control systems can include the following:

Figure 8.11 Exhaust fans are a component of dedicated smoke-control systems. *Courtesy of Texas Fire Prevention Specialists.*

Figure 8.12 Nondedicated smoke-control systems often share components with the HVAC system to move products of combustion.

- Less chance for component failure due to normal use and maintenance
- Lower cost
- Less space needed for mechanical equipment

  The disadvantages of the nondedicated system include:

- Elaborate nature of the system control.
- The possibility of modification of the approved system or smoke controls that might affect the smoke control function.
- All features of the smoke control system may not be exercised during day-to-day operations.

Passive Systems

Passive smoke control is provided by barriers with sufficient fire endurance to provide protection against fire spread. Walls, partitions, floors, doors, and other barriers provide some level of smoke protection to areas that are a dis-

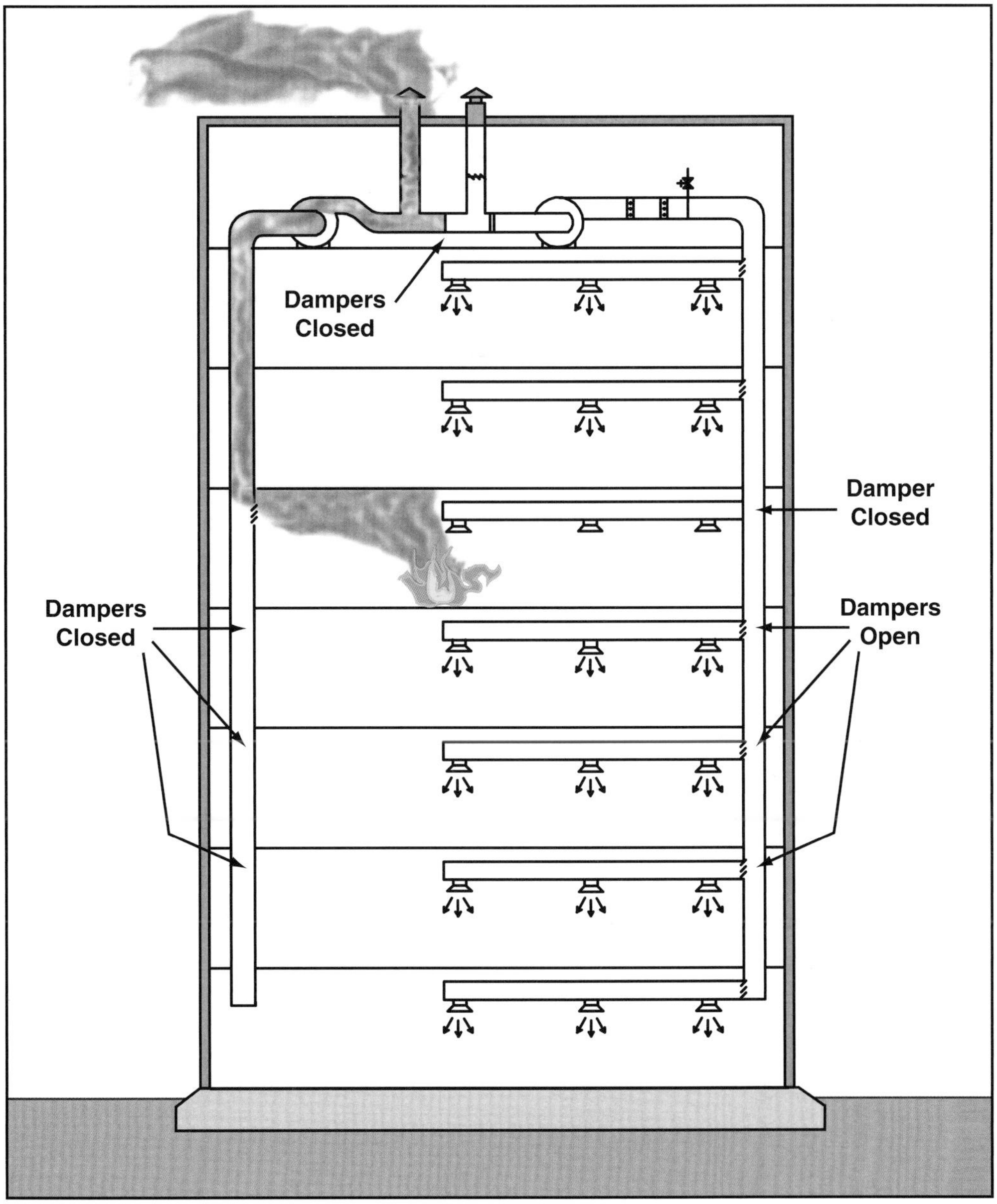

Figure 8.13 A nondedicated smoke-control system operating in the fire mode will exhaust smoke from the fire zone and supply fresh air to adjacent zones.

Labels in figure: Dampers Closed; Dampers Closed; Damper Closed; Dampers Open

tance from the fire's area of origin **(Figure 8.14)**. Barriers that have sufficient fire rating to be effective have long been used to provide protection against smoke and fire spread. Compartmentation provides passive smoke control between defined areas of a building, which are typically referred to as *smoke zones*. Codes, such as NFPA® 101, *Life Safety Code®*, provide criteria for the construction of smoke barriers, including doors and smoke dampers. The amount of protection against smoke leakage depends on the size and shape of the openings in the barriers and the pressure difference across the barriers.

Examples of passive measures include fire stopping of barrier penetrations, door gasket and drop seals, and stair and elevator vestibules **(Figure 8.15)**. Smoke dampers in HVAC ductwork and automatic door devices may also be used to provide smoke control by compartmentation in a given zone. Smoke detectors often trigger smoke dampers and close to block the movement of air through a duct. A sprinkler waterflow switch or another fire alarm component may also trigger dampers. Upon activation, smoke dampers are closed by pneumatic, spring, or electric actuators. Smoke dampers are not designed to prevent the passage of fire, but combination fire/smoke dampers are available if the smoke barrier is also intended to be a fire barrier.

Pressurization Systems

The pressurization method of smoke control uses mechanical fans and ventilation to create a pressure difference across a barrier such as a wall. Pressure differences across a barrier prevent smoke from infiltrating to the high

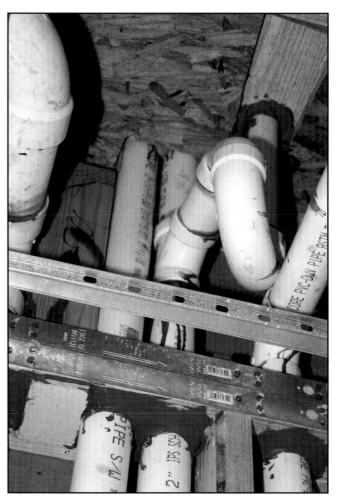

Figure 8.14 Fire doors are a common feature of passive smoke control systems.

Figure 8.15 All penetrations of walls or ceilings should be properly sealed. In this photo, the bottom penetration has been sealed with fire stopping, but the upper penetration has not.

pressure side of the barrier. The airflow through the gaps around a door, construction cracks, and other penetrations do not allow smoke to filter from the low-pressure side which is exposed to smoke from the fire.

Pressurization systems must be designed so that they do not create an excessive pressure that may impede safe egress from a building. If the pressure is too great, it may be difficult to open a door. In a sprinklered facility, the pressure required might be less than what is required for an unsprinklered building.

There are two types of pressurization systems: positive-pressure systems and negative-pressure systems. Positive-pressure systems supply air to the zones adjacent to the zone of the fire's origin. While these systems will adequately contain smoke, they may force smoke from the area of fire origin into unintended areas of the building. In addition, they do not aid in removing smoke from the building.

Negative-pressurization systems serve to exhaust or evacuate the smoke from the area of the fire's origin **(Figure 8.16)**. Shutting down the ventilation system in the areas adjacent to the fire and exhausting smoke from the area of fire origin is the simplest way to accomplish this. With negative-pressure systems, smoke is removed from the building, which improves conditions for both firefighters and occupants.

Stairwell pressurization is a type of positive pressurization that limits the spread of smoke into stairways in high-rise buildings **(Figure 8.17)**. The purpose of these systems is to maintain the integrity of the egress routes for the building's occupants and provide a smoke-free staging area for firefighters. These systems are designed so that there is a pressure difference across a closed stairwell door on the fire floor to prevent infiltration of smoke into the stairwell. Building codes or the authority have jurisdiction (AHJ) may require stairwell pressurization regardless of any other smoke control methods used in the building.

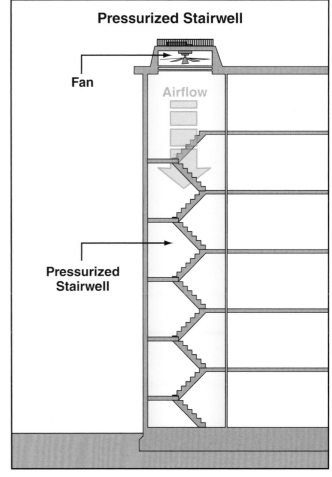

Pressurized Stairwell

Fan

Airflow

Pressurized
Stairwell

Figure 8.16 Negative pressure fans pull products of combustion from the area of fire origin.

Figure 8.17 A pressurized stairwell incorporates a ventilation system that pushes air into the stairwell. The pressure of this air helps keep smoke out of the stairwell.

Pressurization smoke management often depends upon maintaining the integrity of the smoke barriers surrounding the zone to which smoke is to be contained. Maintenance of the integrity of smoke barriers is essential to the operation of the system.

Exhaust Method

The exhaust method is an active smoke-control concept that uses mechanical ventilation along with the properties of smoke to maintain smoke at the highest point in a large space. Because smoke rises to the upper levels of a space, this area serves as a smoke reservoir to contain the smoke **(Figure 8.18)**. A properly designed system should allow the smoke to be maintained at a level of 6 to 10 feet (2 m to 3 m) above the highest occupied floor.

Figure 8.18 The exhaust method uses mechanical ventilation to allow smoke to rise to the upper levels of the space. *Courtesy of the McKinney (TX) Fire Department.*

The most common method of smoke control by exhaust is to provide for exhaust from the upper portions of the space. NFPA® 92 and ASHRAE Standard 52 provide the calculations necessary to design systems for adequate exhaust of large volume spaces. In order to have adequate exhaust, air from outside the compartment must be introduced in a sufficient volume to properly exhaust the compartment. These standards also recognize and describe the use of natural ventilation, although this concept is not widely used in the United States.

The exhaust method is not intended for use in low-ceiling spaces. In addition, sprinkler operation may also adversely impact this approach for spaces with low ceiling heights, because it may cool the smoke, causing it to lose buoyancy.

Opposed Airflow Method

For large openings where pressurization is not applicable, a method of smoke control called *opposed airflow* can be used. This method is typically used for openings in a vertical surface, such as doors or hallways, and where the opening is relatively small in comparison to the size of the surface.

With the opposed airflow method, the following applies:

- Smoke migration from the fire zone is limited by an opposed airflow.

- High velocity air aimed at the area of fire origin keeps the smoke from migrating into unaffected areas.

- To be successful, the air has to be of a sufficient velocity, and the smoke must be diluted and of a relatively low temperature.

- The use of this method does not result in airflow toward the fire, which would intensify the fire or interfere with the egress of the building's occupants.

- In general, the opposed airflow method does not lend itself to use in buildings. However, these systems have been used successfully in other settings such as subway, railroad, and highway tunnels.

NOTE: Use caution for the opposed airflow method due to the concern about supplying oxygen to the fire. The successful use of airflow as a smoke management method depends upon the fire being suppressed or the amount of fuel being restricted to limit the size of the fire.

Dilution

Smoke dilution is sometimes referred to as *smoke purging, smoke removal, smoke exhaust*, or *smoke extraction*. Smoke dilution is most often employed in postfire cleanup but can be used during a fire event.

During a fire event, dilution does the following:

- Reduces the concentration of smoke within a space through supply and exhaust ventilation
- Maintains tenable conditions in a room that is isolated from the fire by smoke barriers and self-closing doors
- Improves conditions within the space when the doors are opened, such as for evacuation or other purposes, and smoke will flow through the doorway

NOTE: This method is not designed to improve conditions within the fire room or areas connected to the fire room.

During a fire, if doors are opened, smoke will flow into areas intended to be protected or safe areas. Supplying outside air into the area can dilute smoke that enters these spaces. This will only be effective in areas remote from the fire. Historically, there has been limited evidence that the use of dilution or smoke purging in the area of the fire will improve conditions in the area or in adjacent spaces.

Zoned Smoke Control

Zoned smoke control is designed to limit the movement of smoke from one compartment of a building to another. With zoned control, a building is divided into a number of smoke zones, separated by partitions and floors. During a fire, mechanical fans are used to contain smoke in the zone of fire origin.

In most situations, each floor of a building is chosen as a separate smoke control zone. However, a zone can be multiple floors or can be a division of a single floor. During a fire, all of the nonsmoke zones or only those adjacent to the fire area may be pressurized. Changes to building codes have eliminated this requirement and now require that **smokeproof enclosures** be used. However, many zoned systems still exist in older buildings.

Zoned smoke control has been used in conjunction with compartmentation. Examples would be hotels or hospitals where the hallways are smoke zones and the rooms are protected by compartmentation.

In some buildings, the HVAC system serves many smoke zones. In order for the system to achieve smoke control, dampers in both the supply and return ducts must be closed as well as the return air damper. For systems where the HVAC system serves only one smoke-control zone, the system has a smoke-zone operational mode and a nonsmoke operational mode.

Smokeproof Enclosures — Stairways that are designed to limit the penetration of smoke, heat, and toxic gases from a fire on a floor of a building into the stairway and that serve as part of a means of egress.

Firefighters' Smoke Control Station (FSCS)

A firefighters' smoke control station (FSCS) provides full monitoring and manual control capability over all smoke control systems and equipment. If manual controls are also provided at other building locations for operation of the smoke control system, the control mode selection from the FSCS should have priority. The smoke control station should be listed for this purpose.

The purpose of the FSCS is to allow firefighters to have control capability over all smoke control system equipment or zones within the building **(Figure 8.19)**. Wherever practical, it is recommended that control be provided by zone, rather than by individual equipment. This will assist firefighters in understanding the operation of the system and will help to avoid problems caused by manually activating equipment in the wrong sequence or by neglecting to control a critical component. The FSCS should be located in the fire command center or other location as approved by the AHJ.

The firefighters' smoke control station (FSCS) should provide the following:

- Ensure that only authorized individuals have access to the FSCS.

- Contain a building diagram that clearly indicates the type and location of all smoke-control equipment as well as the building areas affected by the equipment.

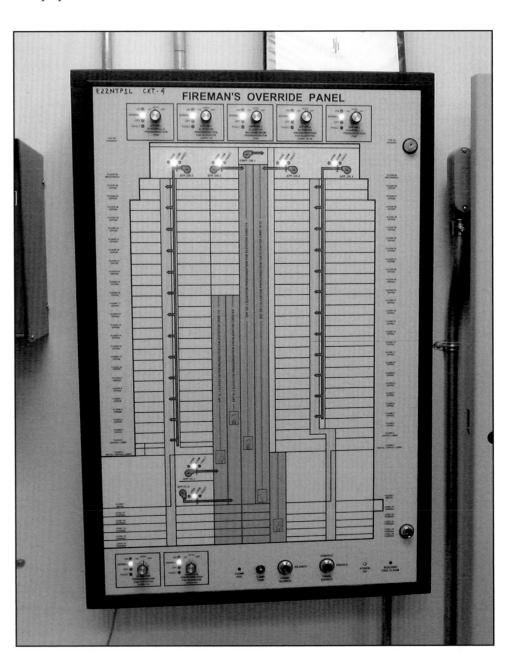

Figure 8.19 As seen on this FSCS, buildings are often divided into zones for smoke control purposes. *Courtesy of Scott Stookey, Austin (TX) Fire Department.*

- Indicate the status of the systems and equipment that are activated.

- Provide manual override switches to restart or shut down the operation of any smoke-control equipment.

- More information about the FSCS can be found in the annexes of NFPA® 92.

System Testing, Inspection, and Maintenance

The three types of testing that are required to ensure the proper operation of smoke management systems over the life of a building include the following:

- Acceptance testing
- Periodic performance testing
- Automatic component testing

Acceptance Testing

The purpose of acceptance testing is to ensure that the final system installation complies with the specified design and that the system is functioning properly. During the planning stages of the project, the building owner and building designer share the criteria for smoke control with the local approving agency. These criteria will include a procedure for acceptance testing. Documents should present a clear understanding of the system, its objectives, and the testing procedures. The AHJ should review and approve these acceptance testing procedures before the final design and acceptance of the system. Prior to the start of the testing process, all building equipment should be in normal operating mode, and wind speed and direction information should be recorded.

A qualified inspector (as approved by the AHJ) should perform acceptance testing of the system to include functional testing and computer modeling where approved. The testing process should include all related fire protection or building systems to the extent that they will affect the operation of the smoke management system. This process includes the testing of smoke detection devices and sprinkler waterflow switches. The inspector must inspect and test the integrity of smoke barriers, partitions, or floor assemblies to verify their operational integrity. This may include pressure, airflow, and leakage testing.

Periodic Performance Testing

Periodic testing is critical to ensure proper operation of a smoke management system. During a periodic test, the operational sequence should be verified and performance of the system rechecked. NFPA® 92 recommends annual or semiannual testing, depending on whether the system is nondedicated or dedicated. It is recommended that dedicated systems be tested semiannually and nondedicated systems be tested annually. A person who is knowledgeable in the operation, testing, and maintenance of the smoke management/control systems should conduct all testing.

Automatic Component Testing

Some systems may be designed or equipped with the ability to perform automatic or self-testing of individual system components. This testing is conducted through the fire alarm or smoke management panel that controls the system. For example, fans may be automatically turned on just long enough to receive positive confirmation of airflow, or dampers may be cycled through an opening and closing process. A report is printed to verify these self-tests, and building personnel should identify any components that fail to operate properly. While this technology is fairly new, it has greatly improved the reliability of smoke management systems in buildings today.

Implications for Emergency Services Personnel

The design and installation of smoke management systems should be included in the plans and design of the building **(Figure 8.20)**. The building office and/ or fire department plans examiner should have knowledge of these systems.

Share the following information with emergency responders within your response area:

- Use prefire planning with specific information about a building's systems and operation.

- Identify the type of smoke control methods, as well as the location of the FSCS and other critical equipment

- Note the operation of the system and the implications on fire suppression operations that might respond to a fire emergency at the location.

- Explain the capabilities and limitations of the particular system to firefighters.

- Consult with the building engineer during preincident planning to help firefighters understand how to use the system to their best advantage during a fire. The building engineer should be available for a consultation as needed during a fire emergency.

- Understand the overall implications of the smoke management system present in the building during fire emergencies.

- Make manual adjustments to the system's operation depending upon the location of the fire, the type of fire, and the location of any occupants.

Figure 8.20 The FSCS and other system controls should be identified and located during preincident planning.

It is imperative that firefighters understand the overall operations, functionality, and power supply for smoke management systems serving buildings in their response area. These systems can make a critical difference in the workload of firefighters and effectiveness of fireground operations, as well as improve survivability of occupants trapped in the building. Some systems may require the primary power to the building be maintained in order for the system to work. Other systems may be supported by standby power systems.

Chapter Summary

Smoke management systems are not a new concept in fire protection. The codes and standards that govern the design, installation, and use of these systems are ever evolving. These systems provide for a safe means of egress for occupants of a building and a safe environment in which firefighters can function. There is a variety of means or methods of providing for fire protection and detection systems, and fire department personnel should become familiar with the types of smoke management systems employed in their area.

Review Questions

1. Name three common hazardous products of combustion.
2. What is the stack effect?
3. What effect do sprinklered buildings have on smoke buoyancy?
4. How can wind affect smoke movement through a building?
5. How can HVAC systems contribute to smoke management through a building?
6. What is the purpose of a smoke management system?
7. How was the Henry Grady Hotel project significant to the history of smoke management systems?
8. What is the difference between dedicated and nondedicated smoke control systems?
9. List different strategies for smoke control.
10. What is the purpose of a firefighters' smoke control station?
11. What three tests should be performed on smoke management systems?
12. What information about smoke management systems should be shared with emergency responders in the response area?

Portable Fire Extinguishers

Chapter Contents

Key Terms

FESHE Outcomes

Fire and Emergency Services Higher Education (FESHE) Outcomes: Fire Protection Systems

11. Explain the operation and appropriate application for the different types of portable fire protection systems.

Portable Fire Extinguishers

Learning Objectives

After reading this chapter, students will be able to:

1. Describe ways that portable fire extinguishers are classified.

2. Explain portable fire extinguisher ratings.

3. Identify the types of extinguishing agents used in portable fire extinguishers.

4. Identify the different methods used to expel extinguishing agents from portable fire extinguishers.

5. Summarize the methods for selection and distribution of portable fire extinguishers.

6. Explain the process for the installation and placement of portable fire extinguishers.

7. Identify criteria for portable fire extinguishers carried on fire apparatus.

8. Explain the inspection, maintenance, and recharging procedures for portable fire extinguishers.

9. Describe hydrostatic testing procedures for portable fire extinguishers.

10. Describe the general techniques for the use of portable fire extinguishers on different classes of fire.

Chapter 9
Portable Fire
Extinguishers

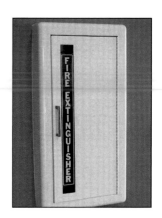

Case History

Following the company barbecue, Jim (the maintenance man) scooped up the coals, put them in a bucket, and set the bucket next to the shelves in the warehouse until he had the time to take it out to the dumpster. A few minutes later, Jim was walking through the warehouse on his way back from the washroom and smelled something burning. He found that the hot coals had set some of the adjacent boxes on fire and the fire was starting to grow. Jim ran for the portable fire extinguisher, but he couldn't remember which aisle it was on. Eventually he found it behind some boxes and made his way back to the fire. Upon pulling the pin and squeezing the trigger, nothing happened. Looking at the extinguisher, Jim saw the pressure gauge reading zero. He ran for another fire extinguisher, found it, and ran back to the fire. This extinguisher was operational, but the fire had grown too large to be extinguished by a portable fire extinguisher. Jim was forced to retreat and call the fire department. The firefighters extinguished the fire, but only after the warehouse sustained significant losses. Jim was treated for minor burns and smoke inhalation. Had the portable fire extinguishers been properly located, adequately marked, and correctly maintained, the incipient fire would have been extinguished by Jim and the losses minimized.

Portable fire extinguishers are some of the most common fire protection devices. They can be found in fixed facilities, such as homes and businesses; in vehicles, including passenger cars; boats; and fire apparatus. Portable fire extinguishers are intended for use on small fires in their incipient or early growth stages. In many cases, a portable extinguisher can control or extinguish a small fire in much less time than it would take for firefighters to deploy a hoseline. In some fire departments, one member on the initial attack team in a high-rise fire carries either a water type or a multipurpose portable extinguisher. If the fire proves to still be small upon reaching it, the extinguisher is used to extinguish the fire.

Many view portable fire extinguishers as the first line of defense against incipient fires. However, a portable fire extinguisher should not be viewed as a substitute for automatic fire protection, detection, and suppression systems. Rather, it should be seen as a complement to these systems.

The value of a portable fire extinguisher lies in the speed with which it can be properly activated and used. An effective fire extinguisher must be:

- Visible and readily accessible **(Figure 9.1, p. 240)**
- Working properly
- Suitable for the hazard being protected

Figure 9.1 Fire extinguishers should be both easily visible and readily accessible.

The purpose of a portable fire extinguisher is to enable an individual with minimal training and orientation to extinguish an incipient fire with reduced risk to the operator. This action must occur after notification of the fire department. As was discussed in Chapter 1, Overview of Fire Protection, Detection, and Suppression Systems, there are many significant fires where notification of the building occupants and the fire department was delayed while portable fire extinguishers were being used. In those instances, the delay resulted in greater damage to life and property.

This chapter provides information on the basic components and types of portable fire extinguishers and explains the ways in which they are classified, rated, tested, inspected, maintained, recharged, and used. Also described in the chapter are common extinguishing agents found in portable fire extinguishers as well as extinguisher selection and placement with respect to fire hazards and fire codes. NFPA® 10, *Standard for Portable Fire Extinguishers*, provides more information on the selection and use of portable fire extinguishers.

Extinguisher Classifications

Portable fire extinguishers are classified based upon the type of fire or fires they are effective at extinguishing. Fires have been broadly grouped into five classifications according to the burning characteristics of various combustible materials. These classifications include:

- **Class A fires** — Involve ordinary combustibles such as wood, cloth, paper, rubber, and many plastics. These fires can be extinguished by cooling, smothering, insulating, or inhibiting the chemical chain reaction.

- **Class B fires** — Involve flammable or combustible liquids and gases, including greases and similar fuels, which can be extinguished by oxygen exclusion, smothering, insulating, and inhibiting the chemical chain reaction.

- **Class C fires** — Involve energized electrical equipment, which requires the use of a nonconductive agent for protection of the extinguisher operator. If electrical power is eliminated, these fires become Class A or Class B, and may be extinguished appropriately.

- **Class D fires** — Involve combustible metals such as magnesium, potassium, sodium, titanium, and zirconium, which require the use of an agent that absorbs heat and does not react with the burning metal.

- **Class K fires** — Involve cooking oils and fats in appreciable depth. With the advent of wet chemical fire suppression systems in commercial kitchens, manufacturers developed K-rated extinguishers. These extinguishers use an identical fire suppression agent as the fixed system. Class K-rated agents work by forming a barrier over the product, thus smothering the fire.

Portable fire extinguishers are labeled according to the classification(s) of fire(s) that they will extinguish. It is critical that the operator be aware of the type of fire and the extinguisher type needed in order to select the proper extinguisher. Symbols are used to identify extinguishers by classification. NFPA® 10 recognizes two different methods of extinguisher recognition: pictorial system and letter-symbol system.

Pictorial System

The pictorial labeling system is the most widely used portable fire extinguisher identification system. The system is designed to make the selection of portable fire extinguishers easier through the use of picture symbols called *pictographs*. Pictographs indicate the type of fire the extinguisher is capable of extinguishing and also indicate when not to use an extinguisher on certain types of fires.

If an extinguisher is suitable for use on a particular fire, the pictograph background is light blue or black. If the extinguisher is not suitable for a particular class of fire, the pictograph symbol has a black background with a diagonal red line through the symbol **(Figure 9.2)**. Some extinguishers are used for more than one classification of fire.

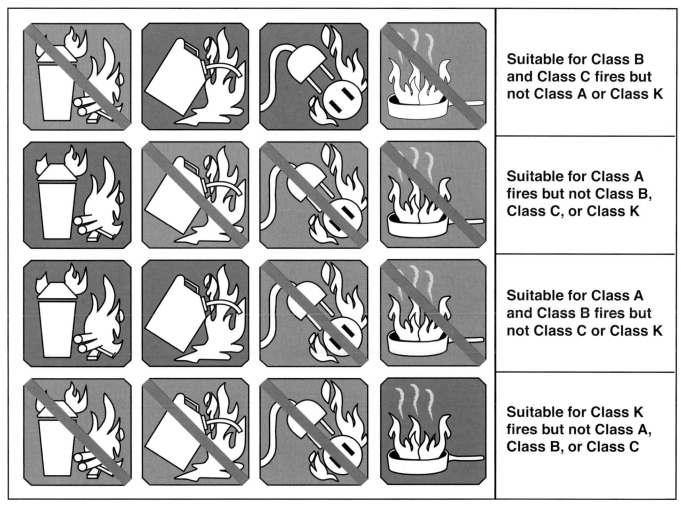

Figure 9.2 These symbols indicate when *not* to use an extinguisher on certain types of fires.

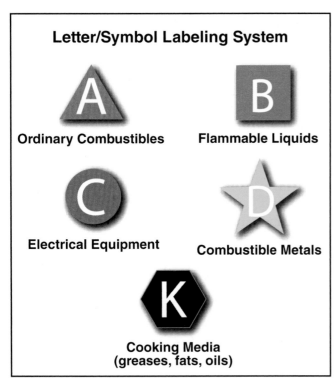

Letter/Symbol Labeling System

A — Ordinary Combustibles

B — Flammable Liquids

C — Electrical Equipment

D — Combustible Metals

K — Cooking Media (greases, fats, oils)

Figure 9.3 The letter-symbol labeling system.

Letter-Symbol Labeling System

The letter-symbol method of extinguisher identification is older than the pictorial system. In the letter-symbol method, each classification of fire is represented by the appropriate letter — A, B, C, D, or K **(Figure 9.3)**. A specific geometric shape encloses the letter. The most recent symbol, the hexagon, is gradually being implemented by most extinguisher manufacturers for Class K applications. In addition, the background of the geometric shape can be color-coded to further identify the extinguisher.

NOTE: The letter-symbol method of portable fire extinguisher identification is rapidly losing its usefulness in favor of the internationally recognized pictograph system.

Portable Fire Extinguisher Ratings

A portable fire extinguisher is rated according to its intended use and fire fighting capability on the five classes of fire. The type and amount of agent contained in the extinguisher and the extinguisher's design determine the amount of fire that can be extinguished for a particular class of fire. This information is conveyed by an alphanumeric classification system designed by Underwriters Laboratories Inc. (UL). NFPA® 10 recommends that this rating information be displayed on the front faceplate of the extinguisher.

The rating system is based upon the extinguishment of test fires in accordance with UL 711, *Standard for Rating and Fire Testing of Fire Extinguishers*. The ratings consist of both a numeric and letter designation for extinguishers intended to combat Class A and Class B fires. Extinguishers classified for Class C fires receive only a letter rating because fires involving energized electrical equipment are fueled by materials that are typically Class A or Class B (or both) in composition. The primary concern in designating an extinguisher for Class C fires is to indicate to its user that it is safe against an electrical shock hazard. Likewise, Class D extinguishers receive only a letter designation. If an extinguisher is capable of extinguishing several classes of fire, the alphanumeric designation will denote this information by showing multiple number-letter ratings on the faceplate.

A 5- to 6-pound (2.5 to 3 kg) dry chemical extinguisher is rated at 2A or 3A depending on its chemical mixture. All other things being equal, a 4A-rated extinguisher should be able to extinguish twice as much fire as a 2A-rated extinguisher. The smallest B:C-type extinguishers have a 5B to 10B rating. All other things being equal, an extinguisher rated 40B should be twice as effective as one rated 20B.

There are several possible combinations of multipurpose extinguishers, not limited to the following:

- A:B
- A:B:C
- B:C **(Figure 9.4)**

NOTE: Class D ratings are never assigned to any type of multipurpose extinguisher.

Portable fire extinguishers are rated by a general criteria. In addition, for each specific classification rating, extinguishers must be subjected to specific tests. These criteria and tests are explained in the sections that follow.

General Rating Criteria

All extinguishers are tested in order to obtain a UL rating. In addition to the fire tests used to determine an extinguisher rating, an extinguisher must also be evaluated on the following criteria:

- Discharge volume capability
- Discharge duration
- Discharge range

Discharge Volume Capability

The following discharge capability characteristics are evaluated:

- Any portable fire extinguisher that uses dry chemical or dry powder as an extinguishing agent must be able to discharge 80 percent of its contents.

- Any portable fire extinguisher that uses agents other than dry chemical or dry powder, such as halon, carbon dioxide, and water, must be able to discharge 95 percent of its contents.

Figure 9.4 This B:C fire extinguisher is suitable for use on both flammable liquid fires and electrical fires.

Discharge Duration

The following discharge duration criteria must be met:

- A portable extinguisher that uses water stored under pressure as an extinguishing agent must have a minimum of a 45- to 60-second discharge time. This time period depends upon the desired rating.

- A minimum effective discharge time is required for any extinguisher for which a Class B rating is desired.

- Dry chemical extinguishers 10 pounds (5 kg) and larger must have a discharge rate of 1 pound per second (lb/sec) (0.5 kg/sec).

Discharge Range

The effective range of the extinguisher stream must meet the following requirements:

- Portable extinguishers that use water stored under pressure must have a minimum effective discharge range of 30 feet (9 m) for a 40-second period.

- Dry chemical and dry powder extinguishers must have a minimum horizontal discharge range of 10 feet (3 m).

Rating Tests

Portable fire extinguishers are classified according to the types of fire for which they are intended. In addition to the classification represented by the letter, Class A and Class B extinguishers are also rated according to performance capability, which is represented by a number. The classification and numerical rating system is based on tests conducted by Underwriters Laboratories Inc. (UL) and Underwriters Laboratories of Canada (ULC). These tests are designed to determine the extinguishing capability for each size and type of extinguisher. **Table 9.1** compares the ratings of each class of portable extinguisher.

Table 9.1 Portable Fire Extinguisher Ratings		
Class	**Ratings**	**Explanations**
A	1-A through 40-A	1-A (1¼ gallons [5 L] of water) 2-A (2½ gallons [10 L] of water)
B	1-B through 640-B	Based on the approximate square foot (square meter) area of a flammable liquid fire a nonexpert can extinguish
C	No extinguishing capability tests	Tests are to determine nonconductivity
D	No numerical ratings	Tested for reactions, toxicity, and metal burnout time
K	No numerical rating	Tested to ensure effectiveness against 2.25 square feet (0.2 m²) of light cooking oil in a deep fat fryer

Fuel Crib — Uniform stacking of wood material where each layer is perpendicular to the layer directly beneath it. Wood is spaced uniformly throughout the crib with separation between the wood material equal to the thickness of the wood and the dimension of the wood is consistent throughout the crib. This is utilized to replicate fire sizes.

Class A Rating Tests

Class A portable fire extinguishers are rated from 1A through 40A. The ratings are based on two different test fires using various sizes of **fuel cribs**. These test fires include a wood crib test and a wood panel test.

Extinguishers rated Class 1A through Class 6A are subjected to both tests. An extinguisher that receives a rating of Class 10A or greater is tested using only the wood crib test **(Figure 9.5)**. Each type of test fire is unique with respect to the configuration and amount of Class A combustibles an extinguisher must extinguish before receiving its rating.

Class B Rating Tests

Like Class A extinguishers, those used for combating Class B fires are classified with a numerical designation. Portable extinguishers suitable for use on Class B fires are classified with a numerical rating ranging from 1B to 640B. The number is an indication of the approximate square foot (ft²) (m²) area of fire involving a 2-inch (50 mm) layer of flammable liquid that can be extinguished by a novice or inexperienced operator. The rating is based on the principle that an expert extinguisher operator, such as a laboratory technician, can extinguish two and one-half times more fire than a novice.

For example, a novice extinguisher operator using a 60B-rated extinguisher can be expected to extinguish a flammable liquid fire involving a 60-square foot (5.5 m²) area. An expert using an extinguisher with the same rating, however, should be able to extinguish a fire involving an area of 150 square feet (13.5 m²). Portable extinguishers with a rating of 20B or greater are considered suitable for outdoor fires.

Class C Rating Tests

A Class C rating is not assigned a numerical designation because the rating signifies only that the extinguishing agent is electrically nonconductive. According to UL 711, the rating is provided in conjunction with Class A, Class B, or Class K ratings, such as 2A:10B:C. No effort is made to indicate the extinguisher's capacity to extinguish a fire that includes energized electrical equipment because fires involving energized electrical equipment are fueled by materials that are Class A, Class B, or Class K in nature. For example, in a Class C fire, it may actually be the insulation material that is burning. The criteria for the rating of a Class C portable extinguisher are established by measuring the electrical conductivity of the extinguisher when it is discharged at an electrically energized target **(Figure 9.6)**.

Figure 9.5 Wood crib tests are used to determine appropriate ratings for fire extinguishers.

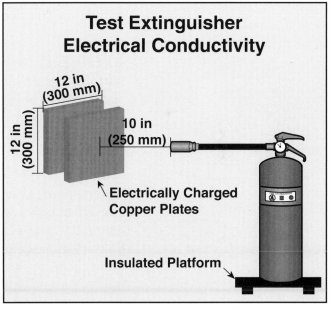

Figure 9.6 A diagram of the test layout for Class C extinguishers.

Class D Rating Tests

Class D portable fire extinguishers are not given numerical ratings. These types of extinguishers are generally tested against fires involving specific combustible metals including magnesium, sodium, and potassium. Other tests, which are different in nature from the standard tests, may be required to address situations or metals where the manufacturer's recommendations or the intended use of the extinguisher is indicated.

In addition to fire tests, each extinguishing agent is evaluated with respect to the adverse effects that might occur in the course of discharge. These effects include agent toxicity, fumes developed, and products of combustion. Also evaluated are the possibilities of adverse reactions resulting from the mistaken use of the agent on a combustible metal.

The following magnesium fire tests are used to evaluate Class D portable fire extinguishers:

- Area fire test
- Pallet-transfer fire test
- Premix fire test
- Casting fire test

For sodium and potassium ratings, two standard procedures are used for testing: the spill-fire test and the pan-fire test. Both tests are conducted when the metals are in a liquid state.

Class K Rating Tests

Class K portable fire extinguishers are rated for their ability to extinguish fire in commercial cooking environments. Tests are performed on a commercial deep fat fryer. The fuel used must be new vegetable shortening or oil with an antifoaming agent and an autoignition temperature of 685°F (360°C) or higher.

Extinguishing Agents

Portable fire extinguishers use many different extinguishing agents. Each agent may be able to control one or more classes of fire, but one agent will not be effective on all classes of fire. This section highlights some of the more common extinguishing agents.

Water

Water is used to extinguish fire by cooling the burning fuel. Water is inexpensive and readily available, and portable water extinguishers — both plain and distilled — are relatively easy to maintain. However, water does have some limitations as an extinguishing agent. The use of water is not as effective as other agents on most Class B fires. Since plain water conducts electricity, it is ineffective and dangerous for use on Class C fires.

Portable water extinguishers have other limitations as well. They are subject to freezing and therefore should be kept in an indoor area. The weight of the water makes these types of extinguishers unwieldy, heavy to transport, and hard to maneuver. A 5-gallon (20 L) water extinguisher weighs in excess of 40 pounds (20 kg) and is the maximum size that is considered truly portable.

Since distilled water has most of the minerals removed, its use is acceptable on Class C fires. This type of portable fire extinguisher uses an atomizing applicator that discharges the water in a very fine mist, contributing to its nonconductive characteristics.

Aqueous Film Forming Foam (AFFF)

Aqueous film forming foam (AFFF) produces both aerated foam and a floating film on the surface of a liquid fuel. AFFF is suitable for both Class A and Class B fires. Most commonly, the AFFF concentrate is premixed with water in the extinguisher and discharged through a special nozzle. Because the foam agent is mixed with water, AFFF is effective on Class A fires by cooling and penetrating the fuel. This agent is very effective on flammable liquid fires because of the double effect of a foam blanket and a surface film to exclude air from the fuel. The AFFF/water mixture has all of the same inherent limitations described for using water as an extinguishing agent.

Another type of AFFF extinguisher contains plain water in a vessel and a special nozzle with a solid form of AFFF concentrate. As the water flows through the nozzle, the concentrate is dissolved in the water to produce a **finished solution**.

NOTE: Some AFFF extinguishers now include an alcohol resistance (AR) rating to identify agents that are effective on polar solvents.

Finished Solution — Extinguishing agent formed by mixing foam concentrate with water and aerating the solution for expansion.

Film Forming Fluoroprotein (FFFP)

Film forming fluoroprotein (FFFP) is a foaming agent that is very similar to AFFF. This agent is usually diluted in a solution of water and foam (3 or 6 percent) and is effective on Class A and Class B fires. These extinguishers are usually located where gasohol and water-soluble flammable liquids are in use.

Carbon Dioxide

Carbon dioxide (CO_2) is a colorless, noncombustible gas that is heavier than air. It extinguishes fire primarily through a smothering action by establishing a gaseous blanket between the fuel and the surrounding air. CO_2 is suitable for Class B and Class C fires. It has very limited value on Class A fires.

CO_2 is stored in the extinguisher in a liquid state, which allows more agent to be stored in a given volume. When discharged from the extinguisher, the CO_2 has a white cloudy appearance, which is due to the small dry ice crystals that may be carried in the gas stream when it is discharged **(Figure 9.7, p. 248)**. Because of its gaseous nature, it is difficult to project it very far from the discharge horn of the extinguisher.

Halons

Halons and other **halogenated agents** contain atoms from one of the halogen series of chemical elements: fluorine, chlorine, bromine, and iodine. The halogenated agents are principally effective on Class B and Class C fires. Halons were originally developed as clean agents because they did not leave any residue when used. However, the original halogenated agents have proven to be harmful to humans and the earth's ozone layer, and through the Montreal Protocol signed in 1987, restrictions phasing out the production of ozone-depleting agents were enacted. For more information on halogenated agents, refer to Chapter 7, Non-Water-Based Fire Suppression Systems.

Halogenated Agents — Chemical compounds (halogenated hydrocarbons) that contain carbon plus one or more elements from the halogen series. Halon 1301 and Halon 1211 are most commonly used as extinguishing agents for Class B and Class C fires.

The main type of halon found in portable extinguishers is bromochlorodifluoromethane, commonly referred to as *Halon 1211*. Bromotrifluoromethane, commonly referred to as *Halon 1301*, is another form that is mainly used in fixed extinguishing equipment.

Halon Replacement Agents

Considerable research and development have been performed on new, clean agents that extinguish fires in the same manner as halon agents, but without the associated damage to the atmosphere. While halon replacement agents have many of the same characteristics of their halon predecessors, they are at a disadvantage due to their greater expense and need for greater quantities of agent to extinguish the fire.

Several categories of clean agents are available such as halocarbon clean agents as well as inert gas clean agents. Halocarbons are either hydrochloro-fluorocarbons (HCFCs) or hydrofluorocarbons (HFCs).

Dry Chemical Agents

Dry Chemical — Any one of a number of powdery extinguishing agents used to extinguish fires. The most common include sodium or potassium bicarbonate, monoammonium phosphate, or potassium chloride.

In physical form, **dry chemical** agents are very small, solid powdery particles **(Figure 9.8)**. Because these particles are solid, they can be projected more effectively from the extinguisher nozzle than gaseous agents. Dry chemicals do not dissipate into the atmosphere as readily as gases and are especially suitable for controlling outdoor fires. The primary disadvantage of using dry chemical portable fire extinguishers is difficulty in cleanup. An airborne dry chemical travels much farther than the immediate fire area, so cleanup can be very time-consuming.

As with all portable fire extinguishers, never refill a dry-chemical type with an agent other than the specific agent for which the extinguisher was designed. Also never mix dry chemical extinguishing agents. Mixing can result in a dangerous chemical reaction, especially with monoammonium phosphate and other dry chemicals.

Figure 9.7 A carbon dioxide extinguisher in use.

Figure 9.8 A dry chemical extinguisher in use.

Several dry chemicals have proven useful as extinguishing agents in portable fire extinguishers. These agents, usually called *ordinary dry chemical agents*, include sodium bicarbonate, potassium bicarbonate, and monoammonium phosphate, which is a multipurpose dry chemical. Urea bicarbonate and potassium chloride are also dry chemical agents but will not be covered in this manual.

Sodium Bicarbonate

Sodium bicarbonate is a form of baking soda and was the first commercially produced dry chemical agent. This agent is still widely used and is treated to be water-repellent and free-flowing. When used as an extinguishing agent, sodium bicarbonate has no toxic effects but may be slightly irritating to the eyes if contact is made. Sodium bicarbonate is color-coded either with blue or white to distinguish it from other dry chemical agents. The agent is effective on Class B and Class C fires.

Sodium bicarbonate has a very rapid knockdown capability against flaming combustion and also has some effect on surface fires in Class A materials. It has been used successfully on textile machinery where fine textile fibers can produce a surface fire.

Potassium Bicarbonate

Potassium bicarbonate, which is also known as *Purple K,* is deemed twice as effective as sodium bicarbonate in fire extinguishment. It is most effective on Class B or Class C fires and is also treated to be water-repellent and free-flowing. Potassium bicarbonate is a purple color to differentiate it from other dry chemicals.

Monoammonium Phosphate

Monoammonium phosphate is an effective and popular agent for use on Class A, Class B, and Class C fires. It has an action similar to other dry chemicals and quickly knocks down flaming combustion. On Class A materials, the monoammonium phosphate forms a solid coating and extinguishes the fire by smothering.

Dry Powder Agents

Dry powder extinguishing agents are those used for extinguishing Class D fires and are not to be confused with dry chemicals **(Figure 9.9)**. Dry powders are designed to extinguish fires involving combustible metals such as aluminum, magnesium, sodium, and potassium. Ordinary extinguishing agents are not capable of controlling fires in combustible metals. Violent reactions may occur, and toxic gases may be released when water contacts the burning metal.

A number of extinguishing agents have been developed for extinguishing fires involving combustible metals. These include some exotic and specialized materials such as foundry flux, trimethoxyboroxine, ternary eutectic chloride, and boron trifluoride. There is no single agent that is effective on all combustible metals. An extinguishing agent must be carefully chosen for the hazard being protected. Three of the more commonly encountered Class D extinguishing agents include Na-X®, Met-L-X®, and Lith-X®.

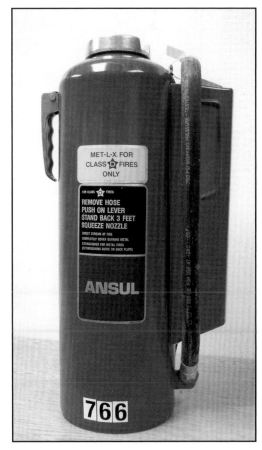

Figure 9.9 This dry powder extinguisher is rated for use on Class D fires.

Na-X®

Na-X® is a Class D extinguishing agent designed specifically for use on sodium, potassium, and sodium-potassium alloy fires. Na-X® is not suitable for use on magnesium fires. Chemically, Na-X® has a sodium carbonate base combined with additives to enhance flow. The extinguishing action forms a crusting or caking on the burning material, causing oxygen-deficiency and thereby extinguishing the fire. Application can be from portable extinguishers or by scoops from pails.

Met-L-X®

Met-L-X® is a sodium chloride-based extinguishing agent intended for use on magnesium, sodium, and potassium fires. Like other dry powders, it contains additives to enhance flowing and prevent caking in the extinguisher. It also extinguishes metal fires by forming a crust on the burning metal to exclude oxygen. The agent is applied from the extinguisher to first control the fire, and then the agent is applied more slowly to bury the fuel in a layer of powder.

Lith-X®

Lith-X® is an agent that can be used on several combustible metals. It was developed to control fires involving lithium but can also be used to extinguish magnesium, zirconium, and sodium fires. Lith-X® consists of a graphite base that extinguishes fires by conducting heat away from the fuel after a layer of the powder has been applied to the fuel. Unlike other dry powders, this agent does not form a crust on the burning material.

Wet Chemical Agents

Wet chemical extinguishing agents are used to suppress fires involving commercial cooking equipment. Most Class K extinguishing agents are alkaline based mixtures consisting of potassium carbonate, potassium acetate, potassium citrate, or a combination. These agents are mixed with water and discharged by an expellant gas. Wet agent is emitted from the extinguisher as a fine mist, which acts to cool the flame front. The misting of the agent also helps to prevent splashing the burning oil.

The agent's primary extinguishing benefit is its ability to mix with the cooking grease to form a foam barrier, or soapy mixture, over the burning fuel through saponification. The Class K solution thus provides a blanketing effect similar to a foam extinguisher but with a greater cooling effect. The saponification process only works on animal fats and vegetable oils.

Types of Portable Fire Extinguishers

Portable fire extinguishers use different methods to expel the extinguishing agent and can be broadly classified according to the method used. These methods include the following:

- Stored-pressure
- Cartridge-operated
- Pump-operated

Obsolete Extinguishers

Fire protection personnel must be alert for portable fire extinguishers that are out of production and no longer suitable for use. Fire service personnel will occasionally encounter such obsolete portable fire extinguishers in old buildings. American manufacturers stopped making inverting-type portable fire extinguishers in 1969. These include soda-acid, foam, internal cartridge-operated water, loaded-stream, and internal-cartridge dry-chemical extinguishers. Manufacturing of extinguishers made of copper or brass with cylinders either soft-soldered or riveted together was also discontinued at this time. Because of the toxicity of carbon tetrachloride and chlorobromomethane, extinguishers using these agents were prohibited in the workplace.

Dry chemical stored-pressure extinguishers produced prior to 1984 must be removed from service at the next maintenance interval. In addition, if any portable fire extinguisher cannot be maintained per manufacturer's specifications, it must be removed from service. These extinguishers should be disposed of according to the operating procedures of the authority having jurisdiction (AHJ).

Stored-Pressure

This type of extinguisher may be found in areas such as office buildings, department stores, or private residences where a high-use factor is not involved. A stored-pressure portable fire extinguisher contains an expellant gas and an extinguishing agent in a single chamber. The pressure of the gas forces the agent out through a siphon tube, valve, and nozzle assembly **(Figures 9.10 a and b)**. Dry chemical extinguishers typically use nitrogen as an expellant gas. In other cases, the expellant gas can be the vapor phase of the agent itself, such as CO_2, in extinguishers. As a highly compressed gas, CO_2 forms its own expellant. Units that use a separate expellant gas have a pressure gauge that permits visual determination as to whether the extinguisher is ready to use.

Stored-pressure extinguishers are easy to use. They usually require only that the operator remove a safety pin, aim the nozzle at the seat or base of the fire, and squeeze the valve handle. However, refilling the unit requires special charging equipment for pressurization, so a certified technician should perform all extinguisher servicing. Many states and municipalities require a license to perform this work.

Stored-Pressure Water Extinguisher

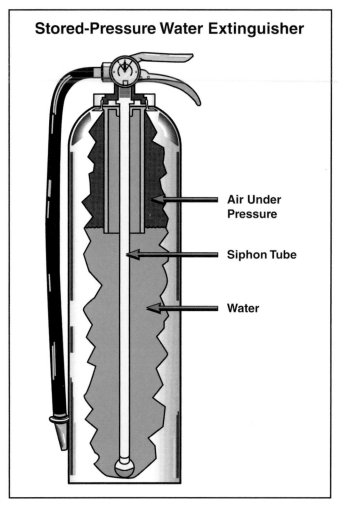

Figure 9.10a In a stored-pressure extinguisher, the expellant gas forces the extinguishing agent out through the siphon tube.

Figure 9.10b A stored-pressure extinguisher features a prominent pressure gauge.

Cartridge-Operated

Cartridge-operated extinguishers are found in industrial operations, such as paint spraying or solvent manufacturing, where they may be used frequently. The cartridge-operated extinguisher has the expellant gas stored in a cartridge, while the extinguishing agent is contained in an adjacent cylinder called an *agent cylinder* or *tank* (**Figure 9.11**). Upon use, the expellant, which is usually CO_2 or nitrogen gas, is released into the agent cylinder. The pressure of the gas forces the agent into the application hose. A handheld nozzle or lever controls the discharge. No pressure gauge is provided.

During inspection, the gas cartridge is weighed to ensure that it has the appropriate amount of propellant. Replacing the gas cartridge and filling the agent cylinder recharges this type of extinguisher. This procedure may be performed in-house and does not require special equipment.

Pump-Operated

A pump-operated extinguisher discharges its agent by the manual operation of a pump and is limited to the use of water as the extinguishing agent. Its primary advantage is that it can be refilled from any available water source in the course of extinguishing a fire. Maintenance is extremely simple and consists mainly of ensuring that the extinguisher is full and has not suffered any mechanical damage.

Selection and Distribution of Portable Fire Extinguishers

Extinguishers must be properly distributed throughout a facility to ensure that they are readily available during an emergency. To be effective, portable extinguishers must be located in close proximity to where they may be needed. In the same manner, extinguishers cannot be effective if there are not enough of them provided for the hazard involved.

Requirements for extinguisher distribution are found in NFPA® 10. These requirements are separated into specifics for Class A, Class B, Class C, Class D, and Class K hazards. Because local codes and ordinances can be more restrictive, these should be reviewed along with the requirements contained in NFPA® 10. Important elements in the selection and distribution of portable fire extinguishers include the following:

- Chemical and physical characteristics of the combustibles that might be ignited
- Potential severity, size, intensity, and rate of advancement of fire
- Location of the extinguisher
- Effectiveness of the extinguisher for the hazard in question
- Personnel available to operate the extinguisher, including their physical abilities and any training they may have in the use of extinguishers
- Environmental conditions that may affect the use of the extinguisher such as temperature, winds, and the presence of toxic fumes or gases
- Any anticipated adverse chemical reactions between the extinguishing agent and the burning material

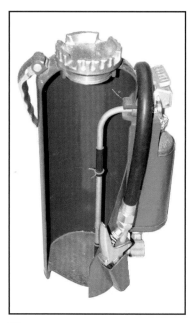

Figure 9.11 A cross section of a cartridge-operated extinguisher.

- Any health and occupational safety concerns such as exposure of the extinguisher operator to heat and products of combustion during fire fighting efforts

- Inspection and service required to maintain the extinguishers

In addition, the type, size, and number of extinguishers needed may vary according to the type of occupancy and the nature of the hazard. NFPA® 10 discusses extinguisher selection based upon the nature of the hazard and the size of the extinguisher needed.

A general method found in NFPA® 10 is used to determine a satisfactory distribution of extinguishers in the vast majority of situations. This method classifies occupancies as light hazard, ordinary hazard, or extra hazard. Portable fire extinguisher distribution is specified on the basis of that classification.

In addition, NFPA® 10 recommends the minimum size of extinguisher and the maximum area to be protected by an extinguisher. Tables contained in NFPA® 10 provide information for the computation of square footage (m²) of a facility coupled with the size of extinguisher required for a minimum number of extinguishers that must be installed. These tables also take into account the hazard classification of the facility.

Class A Factors

In ordinary- or low-hazard occupancies, the AHJ may approve the use of several lower-rated extinguishers in place of one higher-rated extinguisher. For example, two or more extinguishers may be used to satisfy a 6A rating if there are enough individuals trained to use the extinguishers. When the weight of the extinguisher causes problems for those who will be operating it, two extinguishers of lesser weight may be used to replace the heavier extinguisher.

Class B Factors

Flammable liquid fires develop very rapidly and occur in a variety of situations that are fundamentally unique from a fire-control standpoint. When providing extinguishers for Class B hazards, two situations may be encountered. One is a spill fire where the flammable liquid lacks depth. The other involves a flammable liquids fire where the liquid has depth, such as with dip tanks. NFPA® 10 establishes ¼ inch (6 mm) as the criterion for a flammable liquid fire to be classified as having depth. Anything less is considered to be without depth.

Class C and Class D Factors

There are no special spacing rules for Class C hazards because fires involving energized electrical equipment usually involve Class A or Class B fuels. Furthermore, the placement and distribution of portable fire extinguishers for Class D combustible metals cannot be generalized. Determining extinguisher placement involves making an analysis of the specific metal, determining the amount of metal present, determining whether the metal is solid or particulate, and knowing the characteristics of the extinguishing agent **(Figure 9.12, p.254)**.

Figure 9.12 Extinguishers must be tailored to the hazard. In facilities that contain combustible metals, Class D extinguishers must be supplied.

Figure 9.13 Class K extinguishers should be located no more than 30 feet (9 m) from the hazard.

Class K Factors

In the working environment of commercial cooking occupancies, the potential for a hostile fire is always present. Employees in such areas are charged with the responsibility to maintain appropriate cooking temperatures to ensure safety. Employees are usually in various levels of training for their jobs and there is the potential of fire hazards occurring in an assembly area, such as in a dining room. Therefore, NFPA® 10 assigns a more restrictive distance requirement. In areas where Class K fires are likely, the maximum travel distance from the hazard to the extinguisher is 30 feet (9 m) **(Figure 9.13)**.

Installation and Placement of Portable Fire Extinguishers

In addition to proper selection and distribution, the effectiveness of portable fire extinguishers requires that they be readily visible and accessible. Proper extinguisher placement is an essential, but often overlooked aspect of fire protection. Extinguishers should be mounted properly to avoid injury to building occupants and avoid damage to the extinguisher. Some examples of improper mounting would be an extinguisher mounted where it protrudes into a path of travel or one that is sitting on top of a workbench with no mount at all. To minimize these problems, extinguishers are frequently placed in cabinets or wall recesses for protection of both the extinguisher and the people who might walk into them **(Figure 9.14)**. If an extinguisher cabinet is placed in a rated wall, then the cabinet must have the same fire rating as the wall assembly.

Install and place extinguishers so that they are:

- Visible and marked with appropriate signage
- Not blocked by storage or equipment
- Near points of egress or ingress
- Near normal paths of travel

Although it is critical that an extinguisher be properly mounted, it must be placed so that all personnel can access it. For safe lifting, the extinguisher should not be placed too high above the floor. The standard mounting heights specified for extinguishers are as follows **(Figure 9.15)**:

- Extinguishers with a gross weight of 40 pounds (20 kg) or less should be installed so that the top of the extinguisher is not more than 5 feet (1.5 m) above the floor.
- Extinguishers with a gross weight exceeding 40 pounds (20 kg), except wheeled types, should be installed so that the top of the extinguisher is not more than 3½ feet (1 m) above the floor.
- The clearance between the bottom of the extinguisher and the floor should never be less than 4 inches (100 mm).

The physical environment is very important to an extinguisher's reliability, and of greatest concern is the temperature of the environment. These extinguishers must be located where freezing is not possible because testing laboratories evaluate water-based extinguishers at temperatures between 40°F and

Figure 9.14 Extinguishers should be properly mounted to protect both the extinguisher and occupants.

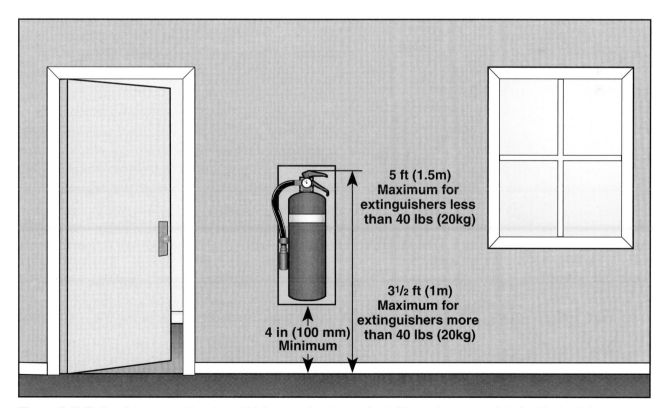

Figure 9.15 Extinguishers must be placed high enough to be easily visible and low enough to be lifted safely.

120°F (4.5°C and 50°C). Other types of extinguishers can be installed where the temperature is as low as -40°F (-40°C). Specialized extinguishers are available for temperatures as low as -65°F (-50°C).

Extinguishers using plain water can be provided with antifreeze recommended by the manufacturer, if necessary. However, exercise care in the use of antifreeze:

- Do not use ethylene glycol in portable fire extinguishers.
- Do not use calcium chloride in stainless steel units.
- Do not add antifreeze to AFFF extinguishers.

Other environmental factors that may adversely affect an extinguisher's effectiveness are corrosive fumes, snow, and rain. A corrosive atmosphere can be encountered not only in an industrial environment but also in marine applications where extinguishers are exposed to saltwater spray. For marine applications, extinguishers are available that have been listed for use in a saltwater environment. In the case of outdoor installations, protect the extinguisher with a plastic bag or place it in a cabinet.

Tampering is also an issue that can affect an extinguisher's ability to operate. Locations, such as college campuses and public facilities, are especially vulnerable to vandalism. Products are commercially available to limit tampering with or removal of extinguishers in these locations.

Portable Fire Extinguishers on Fire Apparatus

Portable fire extinguisher use is not limited to buildings or other structures. In fact, in the hands of a trained firefighter, the extinguisher probably attains its maximum effectiveness. Most of the same considerations that apply to the use of extinguishers by the general public also apply to their use by fire department personnel. Extinguishers used by firefighters must be properly selected, readily accessible, and properly maintained.

Figure 9.16 Extinguishers on fire apparatus should be secured in compartments instead of being exposed to the elements.

Extinguishers carried by the fire department are used more frequently than those in private industry. These extinguishers are subject to harsher use and to a greater variety of environmental conditions and vibrations. To be protected, extinguishers should be carried within compartments rather than in an exposed location **(Figure 9.16)**.

NFPA® 1901, *Standard for Automotive Fire Apparatus*, requires that portable fire extinguishers be carried on apparatus. The extinguishers must be suitable for Class A, Class B, and Class C fires. To date, the NFPA® does not require a Class K extinguisher to be carried on fire apparatus. Although individual portable fire extinguishers are specified by the apparatus purchaser, minimum ratings for the different types of extinguishers to be carried on fire apparatus are as follows:

- One dry chemical extinguisher rated for B:C fires with at least an 80-B:C rating.
- One 2½ gallon (10 L) or larger water extinguisher.

Inspection, Maintenance, and Recharging

Proper servicing is essential to maintaining portable fire extinguisher readiness. The following section highlights the procedures required for properly inspecting, maintaining, and recharging portable fire extinguishers.

Inspection

An inspection is a quick check that visually determines whether a portable fire extinguisher is properly placed and operable **(Figure 9.17)**. A regular inspection ensures that the extinguisher is fully charged and ready to use. Inspections verify that the extinguisher is properly located, is not blocked in any way, and has not been tampered with or emptied. Extinguishers in most occupancies are used infrequently; therefore, this regular inspection is very important to ensure their state of readiness. While this may seem a trivial function, it is critically important because one extinguisher that does not properly operate can result in significant property loss or personal injury.

In order to be effective, inspections must be conducted frequently. NFPA® 10 recommends to perform extinguisher inspections monthly. It is also important to maintain complete records of the inspections. Use inspection tags to record the inspector's name and the date of the inspection. Items to check during a monthly inspection include the following:

- Proper location
- Access (visible and accessible)
- Inspection tag (check for annual inspection)
- Horn or nozzle (look for obstructions)
- Lock pins and tamper seals (ensure they are intact) **(Figure 9.18)**
- Signs of physical damage
- Pressure gauge (ensure extinguisher is fully charged)
- Applicability of extinguisher for hazard classification

Figure 9.17 A regular inspection ensures that the extinguisher is fully charged and ready to use.

Figure 9.18 Check extinguishers to ensure that lock pins and tamper seals are intact.

Maintenance

The purpose of extinguisher maintenance is to ensure that the extinguisher will operate safely. A certified service technician must perform the maintenance and be employed by an extinguisher distribution and/or service company.

Maintenance involves a thorough examination of the mechanical parts, the extinguishing agent, and the expelling means. Maintenance should be performed as required by applicable codes and standards. Nonrechargeable extinguishers should be removed twelve years after the date of manufacture. For more information on portable fire extinguisher maintenance intervals, see NFPA® 10.

Recharging

Recharging is the replacement of the expellant and, if necessary, the agent in a portable fire extinguisher. Recharging is one of the most critical procedures in the maintenance of extinguishers and is not required on a periodic basis for every type of extinguisher.

Recharging some extinguishers involves not only filling the unit with the proper agent, but pressurizing it as well. Pressurization must be performed by properly trained technicians or personnel and with the proper equipment. One obvious danger in pressuring an extinguisher is applying too much pressure to the cylinder. It is important, therefore, to use a source of expellant gas with a pressure not greater than 25 psi (175 kPa) above operating pressure. Another potential danger is the inclusion of moisture in nonwater extinguishers. Moisture can result in the caking of dry chemical, and it also can contribute to interior corrosion and eventual failure of the container.

When refilling an extinguisher, it is important to use only those chemicals or materials specified by the manufacturer or those having an equivalent composition. For example, dry chemicals must be of proper particle size to flow properly. Although sodium bicarbonate is one of the dry chemicals used in extinguishers, the baking soda purchased in a grocery store is not suitable for use in an extinguisher. Extinguishing agents cannot be mixed, and only the extinguishing agent intended for a particular extinguisher can be used in that extinguisher.

Hydrostatic Testing

Hydrostatic Test — A testing method that uses water under pressure to check the integrity of the pressure vessels.

All agent storage cylinders, the discharge hose (if applicable), and the discharge nozzle are required to pass a **hydrostatic test**. This test consists of pressurizing the components to five times their rated capacity for a period of not less than five seconds.

With the exception of pump-tank water extinguishers, portable fire extinguishers are considered pressure vessels. They are either maintained under a constant pressure or are pressurized when they are used. The pressure in extinguishers varies from 100 to 850 psi (700 kPa to 5 950 kPa) depending upon the type. Physical damage and corrosion can cause extinguisher shell failure, which may result in severe injury or death. To ensure that an extinguisher is strong enough to withstand the pressures to which it is subjected, it must be periodically tested.

The method used to pressure test an extinguisher is known as the *hydrostatic test*. Hydrostatic testing consists of filling the cylinder with water and then applying the appropriate pressure by means of a pump. This method is used because it is safer than using a compressed gas. If the cylinder fails while pressurized with water, a violent rupture typically does not occur. Hydrostatic testing of extinguishers is performed at the intervals specified in NFPA® 10 or whenever there is evidence of damage or corrosion. Only trained and experienced personnel should perform hydrostatic testing.

The hydrostatic test pressure is based on the extinguisher's service pressure and its factory test pressure. For CO_2 extinguishers and CO_2 or nitrogen cylinders used as an expellant source, the hydrostatic test pressure is 5/3 or 167 percent of the service pressure stamped on the vessel. The factory test pressure is the pressure at which the vessel was tested at the time of its manufacture. This pressure is shown on the extinguisher nameplate **(Figure 9.19)**. For Halon 1211 and stored-pressure extinguishers, the hydrostatic test pressure is the factory test pressure.

An extinguisher should not be subjected to hydrostatic testing and should be removed from service if it shows signs of physical defects such as damaged threads, corrosion, or welded repairs **(Figure 9.20)**. The following extinguisher vessels that have been burned in a fire should not be hydrostatically tested: a soldered or riveted brass or copper extinguisher vessel and a stainless steel shell that has contained calcium chloride. If an extinguisher vessel ever fails a hydrostatic test, it must be removed from service, made inoperable, and properly destroyed.

As with inspections, records of extinguisher maintenance and hydrostatic testing are important parts of a fire protection program. Extinguisher maintenance that included disassembly is recorded on a plastic collar tag that is

Figure 9.20 Extinguishers should not be subjected to hydrostatic testing if they show signs of damage.

Figure 9.19 The factory test pressure is shown on the extinguisher nameplate.

secured to the neck of the extinguisher handle or other hardware. The information on this tag includes the month and year that the maintenance was performed and the name of the person who performed the work. Hydrostatic test results are recorded on the shell of the extinguisher. Along with the labels attached to the extinguisher, a record-keeping system should be established for management purposes.

Figure 9.21 A firefighter would be expected to extinguish much more fire with an extinguisher than a bystander because of training and experience.

Portable Fire Extinguisher Use

The effectiveness of an extinguisher in dousing a fire depends upon several factors. The most important factor is the user. One person may be able to extinguish a fire while another individual, given the same fire, may not **(Figure 9.21)**. Many extinguishers completely discharge the extinguishing agent in 8 to 15 seconds, which leaves little time for trial and error or practice. Improper use of an extinguisher may injure the operator and may also delay extinguishment.

Upon discovery of any fire, the first action should be the activation of the alarm and notification of the fire agency. In a commercial or industrial occupancy, this action might mean pulling the fire alarm or using a voice paging system to activate an emergency response team. In a smaller environment, such as a residence or small office, this certainly means calling the emergency number, such as 9-1-1, to summon the fire agency. It is critical that this action take place immediately because valuable time is lost when fighting a fire unsuccessfully.

It is important that an extinguisher operator choose the appropriate extinguisher for the fire. The picture-symbol and letter-symbol methods of identification provide a quick means of matching an extinguisher to a classification of fire. This identification is one step that should not be overlooked because choosing the wrong extinguisher could have devastating effects. In many situations, this will not be an issue because extinguishers have been appropriately selected initially for the hazards present in the area or facility. However, individuals should always be encouraged to verify that they have the proper extinguisher for the fire before using it.

The use of a portable fire extinguisher is designed for an untrained person and usually involves three to four basic steps. There are general steps that apply to the use of all extinguishers. These steps are generally referred to as the PASS method. There are also some specific instructional techniques that apply to the use of extinguishers on specific classifications of fire.

PASS Method

Extinguisher operation has been simplified so that even untrained personnel can use them effectively. The acronym PASS serves as a memory aid for the proper operating steps. It is useful as a teaching tool for either firefighters or

Figure 9.22 The PASS application method should be used to place the extinguishing agent on the fire.

the general public. During a fire, every second is of great importance. Therefore, everyone should be acquainted with operating instructions. Steps for the PASS method of extinguishment are as follows **(Figure 9.22)**.

- **P** — Pull the pin at the top of the extinguisher. Break the plastic or thin wire inspection band as the pin is pulled. For cartridge-operated extinguishers, remove the nozzle from its holder, and pressurize the extinguisher by depressing a lever that punctures the cartridge seal.

- **A** — Aim the nozzle or outlet toward the fire. Note that some hose assemblies are clipped to the extinguisher body. The nozzle should be pulled away from the extinguisher and pointed toward the fire.

- **S** — Squeeze the handle above the carrying handle or at the end of the hose to discharge the agent. Give a short test burst to ensure that the extinguisher is operational before approaching the fire. Release the handle to stop the discharge of the agent at any time.

- **S** — Sweep the nozzle back and forth at the base of the fire to disperse the extinguisher agent. Watch for remaining smoldering hot spots or possible reignition of flammable liquids. Ensure that the fire is completely extinguished.

Once the fire has been extinguished, the operator should safely back away from the fire. This safe movement is accomplished by maintaining visual contact with the fire's location and keeping the extinguisher nozzle pointed at the fire in case of reignition. Extinguisher operators should always maintain a means of exit from the fire area and should be prepared to retreat if conditions deteriorate.

NOTE: While the PASS method is appropriate for most extinguishers, some differences exist in the fire attack depending on the extinguishing agent and classification of fire.

Class A Fire Attack

When using a water-based extinguisher, the stream must be aimed at the seat of the fire to maximize the cooling effect of the water on the fuel. Initially, the extinguisher should be used at a distance of 10 to 30 feet (3 to 9 m) from the fire. Attacking a fire from more than 30 feet (9 m) will not be very effective **(Figure 9.23, p. 262)**.

Figure 9.23 Depending on the reach of the extinguisher, a Class A fire should be attacked at a distance of 10 to 30 feet (3 to 9 m).

Figure 9.24 A finger placed over the nozzle orifice can break up the stream pattern if necessary.

When the flames are knocked down, the operator should move closer to wet down any remaining smoldering materials. Fires involving compacted fuels or other deep-seated burning materials must be thoroughly soaked and pulled apart to reach the remaining fire. The extinguisher can be used intermittently to facilitate the soaking. If possible, move the material outside to complete the overhaul process. Place a thumb or finger partially over the nozzle orifice to break the stream into a spray pattern if desired **(Figure 9.24)**.

AFFF and other foam-type extinguishers are effective against Class A fires and are used in a manner similar to water extinguishers. These agents have a low surface tension, which allows the agent to penetrate the fuels, especially tightly packaged fuels.

When using a multipurpose dry chemical extinguisher on a Class A fire, the fire should be attacked at its base while sweeping the nozzle from side to side. Because the multipurpose agent forms a coating on the fuel, it is important that the dry chemical agent thoroughly coats all fuel surfaces. The dry chemical has little cooling effect, and deep-seated fires are difficult to extinguish. If the fire involves lightweight materials that might scatter easily, the dry chemical extinguisher can be used approximately 10 feet (3 m) from and 3 feet (1 m) above the fire. This procedure will cause a cloud of agent to form over the fire and the surrounding area.

CAUTION
Do not inhale the dry chemical agent that is dispersed into the atmosphere as it will irritate the respiratory system.

Class B Fire Attack

Both regular and multipurpose dry chemical extinguishers are used to extinguish fires involving flammable or combustible liquids and gases **(Figure 9.25)**. The extinguishing agent should initially be discharged from a distance of approximately 10 feet (3 m). If the attack is started at a closer range, the velocity of the dry chemical discharge may cause the fuel to splash, thus spreading the fire. In general, a flammable liquid fire should be attacked by sweeping slightly beyond either edge of the fire from a low angle in an attempt not to cause the fuel to splash. This action interrupts the chemical chain reaction and reduces the radiant heat. When fighting flammable liquid fires, it is possible for the fire to flash back across the surface of the liquid. This situation occurs because the extinguishing action of the dry chemical — the interruption of the chemical reaction — is not cumulative like it is with foam. Therefore, an operator must stay alert and be prepared for immediate retreat if a flashback occurs.

If it is necessary to attempt a second attack on the fire, it must be done with a full extinguisher. Flammable liquid fires have the characteristic of being particularly fast spreading. If any doubt exists about the ability to control the fire, personnel should leave the area and await the arrival of the fire department.

CO_2 extinguishers are also effective on flammable liquid fires. Because it is a gas, CO_2 cannot be projected very far out of the nozzle. Therefore, it must be applied at a closer range than other agents. The agent should be applied by sweeping it across the surface to overlap the burning liquid. CO_2 has both a cooling and smothering effect. Consequently, discharge should be continued after initial extinguishment to cool the fuel and prevent flashback.

Hydrocarbon — Organic compound containing only hydrogen and carbon and found primarily in petroleum products and coal.

AFFF is effective on Class B fires involving **hydrocarbons** such as fuel oils, gasoline, and kerosene **(Figure 9.26)**. Without an AR rating, AFFF is not effective on flammable liquids like acetone, alcohols, ethers, and lacquer thinners because they dissolve the foam. AFFF is also not effective on pressurized liquid and gases.

Because AFFF extinguishes fire by establishing a barrier that excludes air and oxygen, the application technique must enhance the formation of the surface film. Maintenance of the foam blanket should be made a priority after extinguishment to prevent reignition. AFFF should not be discharged directly into the liquid surface because it will penetrate the burning liquid and

Figure 9.25 Attacking a Class B fire with a fire extinguisher.

Figure 9.26 AFFF extinguishers are effective on Class B fires.

cause splashing. On fires of depth, the foam should be deflected off the sides or back of the enclosing tank so that the agent will flow down on the liquid. On spill fires, the AFFF can be directed onto the surface just in front of the fire to spray over the fire.

Class C Fire Attack

When attacking fires involving energized electrical equipment, the primary consideration is to ensure that the agent is dielectric or electrically nonconductive so that the operator is not injured. If possible, the equipment should be de-energized before initiating the attack. Whether de-energized or not, fires involving electrical equipment can be fought effectively with halogenated agent, CO_2, dry chemical, or even a water-mist extinguisher. Originally designed to take the place of halon-type portable fire extinguishers, the distilled water portable fire extinguisher has also proven to be effective in extinguishing Class C fires.

CO_2 or water-mist extinguishers are the best choices for use on sensitive electrical equipment. Both agents are nonconductive, noncorrosive, and leave no residue. CO_2 should be applied at close range for a quick knockdown. Water-mist stored-pressure extinguishers can be used from a somewhat greater distance, even as far as 12 feet (3.5 m). The application wand on a water-mist-type extinguisher is designed for operator safety. Halon is effective on Class C fires but is toxic to both the operator and the environment.

Figure 9.27 Class D extinguishing agents are commonly applied with a shovel.

Class D Fire Attack

Various dry powder agents will extinguish fires involving magnesium, sodium, and potassium alloys. Each agent has its limitations, however, and these must be known before attacking a Class D fire. For example, even though a particular agent may work well in combating lithium fires, it may not be effective on magnesium fires. Most dry powder agents extinguish combustible metals by caking and adhering to the material, which excludes the necessary air required for combustion. The burning material should be covered with a 2-inch (50 mm) layer of dry powder agent. Dry powder agents should be applied in a manner that minimally disturbs the burning material. Most are applied with either an extinguisher or a shovel **(Figure 9.27)**.

Class K Fire Attack

Class K fires are particularly difficult to extinguish because of their tendency to reignite. Although a Class K fire may have been extinguished properly with a dry chemical, the fuel can change chemically and reach autoreignition at a lower temperature. For this reason, only an extinguisher with a Class K rating is recommended for use on this type of fire. A Class A:B:C- or Class B:C-rated portable fire extinguisher may indeed work effectively, but the operator should pay attention for a possible autoreignition and a second or third attack on the fire.

The initial attack on a fire with a Class K extinguisher is similar to other types of attack. Application of the agent should begin from a distance of 10 to 12 feet (3 to 3.5 m) from the burning material. The operator should hold the application wand at the edge of the flames and then coat the surface of the material with a side-to-side sweep. This application should continue until the portable fire extinguisher is completely empty. Fire extinguishment takes place through the removal of oxygen and a partial cooling of the fuel. It is the innate cooling quality of the agent, along with its ability to form soapy foam or saponify, that prevents the fuel from reaching a lower reignition temperature.

Chapter Summary

Portable fire extinguishers have long been recognized and used as a first line of defense against incipient fires. When extinguishers are correctly selected and placed, inspected, maintained, and used with proper training, they have proven to save lives and property.

Not all hazards are alike and therefore not all portable extinguishers are alike. In addition, extinguishers are useless if they are unavailable, not suited for the particular fire, or nonfunctioning. Personnel responsible for selection and placement of the correct extinguisher in a facility are critical components to fire and life safety. These personnel require knowledge of the operating principles of various extinguishers, the advantages and disadvantages of different extinguishing agents, and familiarity with inspection and maintenance procedures. Personnel must also be alert for obsolete or damaged extinguishers so that they can be removed from service before they cause injury when someone attempts to use them. Finally, individuals in a workplace need to be trained in the basics of extinguisher operation so that they may be able to control a fire and evacuate the premises until additional help arrives.

Review Questions

1. What are the classifications of portable fire extinguishers?
2. What is the most widely used symbol system on portable fire extinguishers?
3. What factors are the ratings of portable fire extinguishers based upon?
4. What testing criteria are used to rate portable fire extinguishers?
5. Why are some fire extinguishers not given numerical ratings?
6. What types of extinguishing agents are used in portable fire extinguishers?
7. List the methods used to expel extinguishing agents.
8. What elements are important in the selection and distribution of portable fire extinguishers?
9. What are factors to consider when properly placing a portable fire extinguisher?
10. What classes of extinguishers must be carried on fire apparatus?

11. List some of the items that should be checked during a monthly inspection.

12. What are the potential dangers involved in recharging portable fire extinguishers?

13. When should a fire extinguisher not be subjected to hydrostatic testing and be removed from service?

14. Explain the PASS method.

15. Differentiate between the five classes of fire attack.

Appendix A
Chapter and Page Correlation to FESHE Requirements

FESHE Course Outcomes	Chapter References	Page References
1. Explain the benefits of fire protection systems in various types of structures.	1, 4, 5, 7, 8	12-16, 20-21, 106-112, 134, 153-157, 201-212, 224-233
2. Describe the basic elements of a public water supply system including sources, distribution networks, piping, and hydrants.	3	83-86, 89-94, 96-99
3. Explain why water is a commonly used extinguishing agent.	3	70-75
4. Identify the different types and components of sprinkler, standpipe, and foam systems.	4, 5, 6, 7	107-134, 137-143, 152-157, 168-189, 201-212
5. Review residential and commercial sprinkler legislation.	1	18-20
6. Identify the different types of non-water-based fire suppression systems.	7	201-212
7. Explain the basic components of a fire alarm system.	2	27-31
8. Identify the different types of detectors and explain how they detect fire.	2	32-57
9. Describe the hazards of smoke and list the four factors that can influence smoke movement in a building.	8	218-223
10. Discuss the appropriate application of fire protection systems.	4, 5, 6, 7, 8	106-112, 127-134, 137-140, 153-156, 169-182, 188-189, 201-212, 224-235
11. Explain the operation and appropriate application for the different types of portable fire protection systems.	9	240-246, 250-256, 260-265

Glossary

A

Acceptance Test — Preservice test on fire detection and/or suppression systems after installation to ensure that the system operates as intended.

Alarm Signal — Signal given by a fire detection and alarm system when there is a fire condition detected.

Atmospheric Pressure — Force exerted by the atmosphere at the surface of the earth due to the weight of air. Atmospheric pressure at sea level is about 14.7 psi (101 kPa). Atmospheric pressure increases as elevation is decreased below sea level and decreases as elevation increases above sea level.

Automatic Sprinkler System — System of water pipes, discharge nozzles, and control valves designed to activate during fires by automatically discharging enough water to control or extinguish a fire. *Also known as* Sprinkler System.

B

Bimetallic — Strip or disk composed of two different metals that are bonded together; used in heat-detection equipment.

British Thermal Unit (Btu) — Amount of heat energy required to raise the temperature of 1 pound of water 1°F. 1 Btu = 1.055 kilojoules (kJ).

C

Cavitation — Condition in which vacuum pockets form due to localized regions of low pressure at the vanes in the impeller of a centrifugal pump and causes vibrations, loss of efficiency, and possibly damage to the impeller.

Central Station System — Alarm system that functions through a constantly attended location (central station) operated by an alarm company. Alarm signals from the protected property are received in the central station, and trained personnel then transmit them to the fire department alarm communications center.

Centrifugal Pump — Pump with one or more impellers that rotate and utilize centrifugal force to move the water. Most modern fire pumps are of this type.

Check Valve — Automatic valve that permits liquid flow in only one direction. For example, the inline valve that prevents water from flowing into a foam concentrate container when the nozzle is turned off or there is a kink in the hoseline.

Churn — Rotation of a centrifugal pump impeller when no discharge ports are open so that no water flows through the pump.

Circulating Feed — Fire hydrant that receives water from two or more directions.

Circulation Relief Valve — Small relief valve that opens and provides enough waterflow into and out of the pump to prevent the pump from overheating when it is operating at churn against a closed system.

Clean Agent Fire Suppression System — System that uses special extinguishing agents that leave little or no residue.

Code — A collection of rules and regulations enacted by a legislative body to become law in a particular jurisdiction.

Consensus Standard — Rules, principles, or measures that are established though agreement of members of the standards-setting organization.

D

Dead-End Main — Water main that is not looped and in which water can flow in only one direction.

Driver — Engine or motor used to turn a pump.

Dry-Barrel Hydrant — Operating valve underground at the base of the hydrant rather than in the aboveground portion of the hydrant barrel. The valve actually is in the hydrant barrel and not at the water main. When operating properly, there is no water in the barrel of the hydrant when it is not in use. These hydrants should be used in areas where freezing could occur.

Dry Chemical — Any one of a number of powdery extinguishing agents used to extinguish fires. The most common include sodium or potassium bicarbonate, monoammonium phosphate, or potassium chloride.

Dry Chemical Fire Suppression System — System that uses dry chemical powder as the primary extinguishing agent; often used to protect areas containing volatile flammable liquids.

E

Elevation Head — Pressure related to the difference in elevation between the water supply and the discharge orifice, commonly expressed in feet.

F

Finished Solution — Extinguishing agent formed by mixing foam concentrate with water and aerating the solution for expansion.

Fire Alarm Control Unit — The main fire alarm system component that monitors equipment and circuits, receives input signals from initiating devices, activates notification appliances, and transmits signals off-site.

Fire Detection System — System of detection devices, wiring, and supervisory equipment used for detecting fire or products of combustion and then signaling that these elements are present.

Fire-Gas Detector — Device used to detect gases produced by a fire within a confined space.

Fire Pump — Water pump used in public and private fire protection to provide water supply to installed fire protection systems.

Fire Suppression System — System designed to act directly upon the hazard to mitigate or eliminate it, not simply to detect its presence and/or initiate an alarm.

Fixed-Temperature Heat Detector — Temperature-sensitive device that senses temperature changes and sounds an alarm at a specific point, usually 135°F (57°C) or higher.

Flame Detector — Detection device used in some fire detection systems (generally in high-hazard areas) that detect light/flames in the ultraviolet wave spectrum (UV detectors) or detect light in the infrared wave spectrum (IR detectors).

Flow Pressure — Pressure created by the rate of flow or velocity of water coming from a discharge opening.

Friction Loss — That part of the total pressure lost as water moves through a hose or piping system; caused by water turbulence and the roughness of interior surfaces of hose or pipe.

Fuel Crib — Uniform stacking of wood material where each layer is perpendicular to the layer directly beneath it. Wood is spaced uniformly throughout the crib with separation between the wood material equal to the thickness of the wood and the dimension of the wood is consistent throughout the crib. This is utilized to replicate fire sizes.

Fusible Element — Dissimilar metals that fuse or melt when exposed to heat and allows the circuits to open or close and transmit a signal to a FACU; commonly found in fixed temperature heat detectors.

G

Governor — Built-in pressure regulating device to control pump-discharge pressure by limiting engine rpm.

Grid System

Grid System — Interconnecting system of water mains in a crisscross or rectangular pattern.

H

Halogenated Agents — Chemical compounds (halogenated hydrocarbons) that contain carbon plus one or more elements from the halogen series. Halon 1301 and Halon 1211 are most commonly used as extinguishing agents for Class B and Class C fires.

Halon — Halogenated agent; extinguishes fire by inhibiting the chemical reaction between fuel and oxygen.

Head — Alternate term for pressure, especially pressure due to elevation. For every 1-foot increase in elevation, 0.434 psi is gained (for every 1-meter increase in elevation, 9.82 kPa is gained). *Also called* Head Pressure.

Horizontal Split-Case Centrifugal Pump — Centrifugal pump with the impeller shaft installed horizontally and often referred to as a split-case pump. This is because the case in which the shaft and impeller rotates is split in the middle and can be separated, exposing the shaft, bearings, and impeller.

House Line — Permanently fixed, private standpipe hoseline.

Hydrocarbon — Organic compound containing only hydrogen and carbon and found primarily in petroleum products and coal.

Hydrostatic Test — A testing method that uses water under pressure to check the integrity of the pressure vessels.

I

Impeller — Vaned, circulating member of the centrifugal pump that transmits motion to the water.

Indicating Valve — Water main valve that visually shows the open or closed status of the valve.

Ionization Detector — Type of smoke detector that uses a small amount of radioactive material to make the air within a sensing chamber conduct electricity.

K

K-Factor — Discharge coefficient used in hydraulic calculations of sprinkler systems and is based upon the sprinkler discharge orifice and physical characteristics of the sprinkler.

L

Latent Heat of Vaporization — Quantity of heat absorbed by a substance at the point at which it changes from a liquid to a vapor.

Loop System — Water main arranged in a complete circuit so that water will be supplied to a given point from more than one direction.

M

Manual Pull Station — Manual fire alarm activator.

Mass Notification System (MNS) — System that notifies occupants of a dangerous situation and allows for information and instructions to be provided.

Mitigate — To cause to become less harsh or hostile; to make less severe, intense, or painful; to alleviate.

Multiple-Stage Pumps — Any centrifugal fire pump having more than one impeller.

P

Passive Smoke Control — Smoke control strategies that incorporate fixed components that provide protection against the spread of smoke and fire. Passive smoke control components include fire doors, fire walls, fire stopping of barrier penetrations, and stair and elevator vestibules.

Photoelectric Smoke Detector — Type of smoke detector that uses a small light source, either an incandescent bulb or a light-emitting diode (LED), to detect smoke by shining light through the detector's chamber: smoke particles reflect the light into a light-sensitive device called a photocell.

Pipe Schedule — Thickness of the wall of a pipe.

Positive-Displacement Pump — Self-priming pump that utilizes a piston or interlocking rotors to move a given amount of fluid through the pump chamber with each stroke of the piston or rotation of the rotors.

Post Indicator Valve (PIV) — A type of valve used to control underground water mains that provides a visual means for indicating "open" or "shut" positions; found on the supply main of installed fire protection systems.

Predischarge Warning Device — Alarm that sounds before a total flooding fire extinguishing system is about to discharge. This gives occupants the opportunity to leave the area.

Pressure — Force per unit area exerted by a liquid or gas measured in pounds per square inch (psi) or kilopascals (kPa).

Pressure Maintenance Pump — A pump used to maintain pressure on a fire protection system in order to prevent false starts at the fire pump.

Pressure Relief Valve — Pressure control device designed to eliminate hazardous conditions resulting from excessive pressures by allowing this pressure to bypass to the intake side of the pump.

Priming — To create a vacuum in a pump by removing air from the pump housing and intake hose in preparation to permit the drafting of water.

Proprietary Alarm System — Fire protection system owned and operated by the property owner.

Protected Premises System — Alarm system that alerts and notifies only occupants on the premises of the existence of a fire so that they can safely exit the building and call the fire department. If a response by a public safety agency (police or fire department) is required, an occupant hearing the alarm must notify the agency.

Public Emergency Alarm Reporting System — System that connects the protected property with the fire department alarm communications center by a municipal master fire alarm box or over a dedicated telephone line.

Pump Controller — Electric control panel used to switch a fire pump on and off and to control its operation.

Pyrolysis — Thermal or chemical decomposition of fuel (matter) because of heat that generally results in the lowered ignition temperature of the material.

R

Rate-of-Rise Heat Detector — Temperature-sensitive device that sounds an alarm when the temperature changes at a preset value, such as 12°F to 15°F (7°C to 8°C) per minute.

Reducer — Adapter used to attach a smaller hose to a larger hose. The female end has the larger threads, while the male end has the smaller threads.

Remote Receiving System — System in which alarm signals from the protected premises are transmitted over a leased telephone line to a remote receiving station with a 24-hour staff; usually the municipal fire department's alarm communications center.

Residual Pressure — Pressure measured at the hydrant to which a pressure gauge is attached while water is flowing from one or more other hydrants during a hydrant flow test. It represents the pressure remaining in the water supply system while the test water is flowing and is that part of the total pressure that is not used to overcome friction or gravity while forcing water through fire hose, pipe, fittings, and adapters.

Retard Chamber — A collection chamber used on wet systems with alarm check valves whose function is to prevent false alarms due to water pressure surges.

Riser — Vertical water pipe used to carry water for fire protection systems aboveground, such as a standpipe riser or sprinkler riser.

S

Saponification — A phenomenon that occurs when mixtures of alkaline-based chemicals and certain cooking oils come into contact, resulting in the formation of a soapy film.

Service Test — Series of tests performed on fire detection and/or suppression systems in order to ensure operational readiness. These tests should be performed at least yearly or whenever the system has undergone extensive repair or modification.

Single-Stage Pump — Centrifugal pump with only one impeller.

Smoke Control System — Engineered systems designed to control smoke by the use of mechanical fans to produce airflows and pressure differences across smoke barriers to limit and direct smoke movement.

Smoke Damper — Device that restricts the flow of smoke through an air-handling system; usually activated by the building's fire alarm signaling system.

Smoke Detector — Alarm-initiating device designed to actuate when visible or invisible products of combustion (other than fire gases) are present in the room or space where the unit is installed.

Smoke Management System — System that limits the exposure of building occupants to smoke. May include a combination of compartmentation, natural or mechanical means to control ventilation and smoke migration from the affected area, as well as a means of removing smoke to the exterior of the building.

Smokeproof Enclosures — Stairways that are designed to limit the penetration of smoke, heat, and toxic gases from a fire on a floor of a building into the stairway and that serve as part of a means of egress.

Specific Gravity — Weight of a substance compared to the weight of an equal volume of water at a given temperature. A specific gravity less than 1 indicates a substance lighter than water; a specific gravity greater than 1 indicates a substance heavier than water.

Specific Heat — The amount of heat required to raise the temperature of a specified quantity of a material and the amount of heat necessary to raise the temperature of an identical amount of water by the same number of degrees.

Sprinkler — Waterflow discharge device in a sprinkler system; consists of a threaded intake nipple, a discharge orifice, a heat-actuated plug, and a deflector that creates an effective fire stream pattern that is suitable for fire control.

Stack Effect — Phenomenon of a strong air draft moving from ground level to the roof level of a building. The air movement is affected by building height, configuration, and temperature differences between inside and outside air.

Standpipe System — Wet or dry system of pipes in a large single-story or multistory building with fire hose outlets installed in different areas or on different levels of a building to be used by firefighters and/or building occupants. The system is used to provide for quick deployment of hoselines during fire fighting operations.

Static Pressure — (1) Potential energy that is available to force water through pipes and fittings, fire hose, and adapters. (2) Pressure at a given point in a water system when no testing or fire protection water is flowing.

Stratification — Formation of smoke into layers as a result of differences in density with respect to height with low density layers on the top and high density layers on the bottom.

Supervisory Signal — Signal given by a fixed fire protection system when there is a condition in the system that is off-normal.

T

Tenable Atmosphere — Capable of maintaining human life.

Thermal Element — Device used in sprinklers and some fire detection equipment that is designed to activate when temperatures reach a predetermined level.

Transcription — Method by which an AHJ adopts a code in whole to become a new regulation.

Trouble Signal — Signal given by a fixed fire protection alerting system when a power failure or other system malfunction occurs.

V

Vertically Mounted Split-Case Centrifugal Pump — Centrifugal pump similar to the horizontal split-case, except that the shaft is oriented vertically and the driver is mounted on top of the pump.

W

Waterflow Device — Initiating device that recognizes movement of water within the sprinkler or standpipe system. Once movement is noted, the waterflow device activates a local alarm and/or may transmit a signal to the FACU.

Wet Barrel Hydrant — Fire hydrant that has water all the way up to the discharge outlets. The hydrant may have separate valves for each discharge or one valve for all the discharges. This type of hydrant is only used in areas where there is no danger of freezing weather conditions.

Wet Chemical Fire Suppression System — Suppression system that uses a wet chemical solution as the primary extinguishing agent; usually installed in range hoods and associated ducting where grease may accumulate.

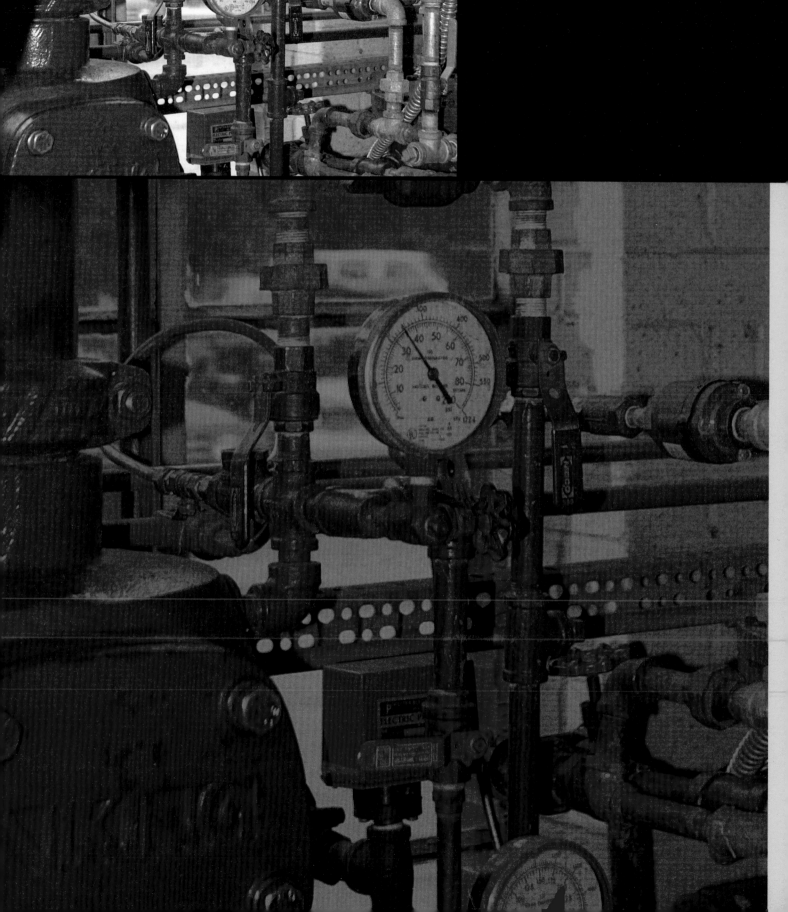

Index

Indexed by Nancy Kopper.